ADVANCES IN MARINE NAVIGATION AND SAFETY OF SEA TRANSPORTATION

T0239625

Advances in Marine Navigation and Safety of Sea Transportation

Editors

Adam Weintrit & Tomasz Neumann
Gdynia Maritime University, Gdynia, Poland

CRC Press
Taylor & Francis Group
Boca Raton London New York

CRC Press is an imprint of the
Taylor & Francis Group, an **informa** business

A BALKEMA BOOK

Published by:
CRC Press/Balkema
P.O. Box 447, 2300 AK Leiden, The Netherlands
e-mail: Pub.NL@taylorandfrancis.com
www.crcpress.com – www.taylorandfrancis.com

First issued in paperback 2020

© 2019 by Taylor & Francis Group, LLC
CRC Press/Balkema is an imprint of the Taylor & Francis Group, an informa business

No claim to original U.S. Government works

Typeset by MPS Limited, Chennai, India

ISBN 13: 978-0-367-72853-3 (pbk)
ISBN 13: 978-0-367-35760-3 (hbk)

Visit the Taylor & Francis Web site at
http://www.taylorandfrancis.com

and the CRC Press Web site at
http://www.crcpress.com

Contents

List of reviewers

*Prof. Nicoleta **Acomi**, Constanta Maritime University, Constanta, Romania*
*Prof. Sauli **Ahvenjärvi**, Satakunta University of Applied Sciences, Rauma, Finland*
*Prof. Antonio **Angrisano**, Università Telematica 'Giustino Fortunato', Benevento, Italy*
*Prof. Andrzej **Bąk**, Maritime University of Szczecin, Poland*
*Prof. Sanja **Bauk**, University of Montenegro, Kotor, Montenegro*
*Prof. Tor Einar **Berg**, SINTEF Ocean, Trondheim, Norway*
*Prof. Shwu-Jing **Chang**, National Taiwan Ocean University (NTOU), Keelung, Taiwan*
*Prof. Marcin **Chrzan**, University of Technology and Humanities, Radom, Poland*
*Prof. Stephen J. **Cross**, International Marine Simulator Forum, Global Maritime Academics, The Netherlands*
*Prof. Kevin **Cullinane**, University of Gothenburg, Sweden*
*Prof. Patrick **Donner**, World Maritime University, Malmö, Sweden*
*Prof. Eamonn **Doyle**, National Maritime College of Ireland, Cork, Ireland - Emeritus Professor*
*Prof. Andrzej **Felski**, Polish Naval Academy, Gdynia, Poland*
*Prof. Alberto **Francescutto**, University of Trieste, Trieste, Italy*
*Prof. Wieslaw **Galor**, Maritime University of Szczecin, Poland*
*Prof. Hassan **Ghassemi**, Amirkabir University of Technology (AUT), Teheran, Iran*
*Prof. Witold **Gierusz**, Gdynia Maritime University, Gdynia, Poland*
*Prof. Floris **Goerlandt**, Dalhousie University, Halifax, Canada*
*Prof. Lucjan **Gucma**, Maritime University of Szczecin, Poland*
*Prof. Maciej **Gucma**, Maritime University of Szczecin, Poland*
*Prof. Gregory **Hanchrow**, Maritime College, State University of New York (SUNY), New York, The United States*
*Prof. Kalin **Kalinov**, Nikola Y. Vaptsarov Naval Academy, Varna, Bulgaria*
*Prof. Witold **Kazimierski**, Maritime University of Szczecin, Poland*
*Prof. Lech **Kobylinski**, The Ship Handling Research and Training Centre at Ilawa, Poland*
*Prof. Vitaliy **Koshevyy**, Odessa National Maritime Academy, Odessa, Ukraine*
*Prof. Przemysław **Krata**, Gdynia Maritime University, Poland*
*Prof. Serdar **Kum**, Istanbul Technical University, Istanbul, Turkey*
*Prof. Artur **Makar**, Polish Naval Academy, Gdynia, Poland*
*Prof. Jerzy **Matusiak**, Helsinki University of Technology, Helsinki, Finland*
*Prof. Jerzy **Mikulski**, University of Economics, Katowice, Poland*
*Prof. Waldemar **Mironiuk**, Polish Naval Academy, Gdynia, Poland*
*Prof. Nikitas **Nikitakos**, University of the Aegean, Chios, Greece*
*Prof. Andy **Norris**, The Royal Institute of Navigation, London, The United Kingdom*
*Prof. Zbigniew **Otremba**, Gdynia Maritime University, Gdynia, Poland*
*Prof. Vytautas **Paulauskas**, Klaipeda University, Klaipeda, Lithuania*
*Prof. Marzenna **Popek**, Gdynia Maritime University, Gdynia, Poland*
*Prof. Thomas T. **Porathe**, Norwegian University of Science and Technology, Trondheim, Norway*
*Prof. Malek **Pourzanjani**, South African International Maritime Institute (SAIMI), Port Elizabeth, South Africa*
*Prof. Dorota **Pyć**, University of Gdańsk, Poland*
*Prof. Martin **Renilson**, Australian Maritime College, University of Tasmania, Launceston, Australia*
*Prof. Helen **Sampson**, Cardiff School of Social Sciences, Cardiff University, The United Kingdom*
*Prof. Arnaud **Serry**, Université du Havre, Le Havre, France*
*Prof. Sutanto **Soehodho**, University of Indonesia, Jakarta, Indonesia*
*Prof. Andrzej **Stateczny**, Gdańsk University of Technology, Gdańsk, Poland*
*Prof. Francisco **Taveira-Pinto**, University of Porto (FEUP), Porto, Portugal*
*Prof. Vladimir **Torskiy**, Odessa National Maritime Academy, Odessa, Ukraine*
*Prof. Janusz **Uriasz**, Maritime University of Szczecin, Poland*
*Prof. Byeong-Deok **Yea**, Korea Maritime and Ocean University, Busan, Korea*
*Prof. Homayoun **Yousefi**, Chabahar Maritime University, Chabahar, Iran*

Advances in Marine Navigation and Safety of Sea Transportation. Introduction

A. Weintrit & T. Neumann
Gdynia Maritime University, Gdynia, Poland
Poland Branch of the Nautical Institute

The contents of the book are partitioned into four separate chapters: Chapter 1 - Advances in Marine Navigation; Chapter 2 - Inland, Costal, Port and Offshore Innovations; Chapter 3 - New Perspective for Communication Systems; Chapter 4 - Challenges in Safety of Sea Transportation.

In each of them readers can find a few subchapters. Subchapters collected in the first chapter, titled 'Advances in Marine Navigation', concern the following aspects: analysis of primary position validation in ECDIS, comparison between simulator modelled and empirical ship squat prediction, environmental impact of new maritime routes Cadiz-Huelva-Algarve, optimal route planning algorithm for coastal ships considering ocean environment and grounding, probabilistic modelling of ship-bank contacts based on manoeuvring performance under environmental loads, and statistical analysis of the real surface currents and wind parameters for the Szczecin Lagoon.

In the second chapter there are described problems related to inland, costal, port and offshore innovations covering: a novel algorithm for modelling human decision making of inbound merchant ships (a case study of the Shanghai Waigaoqiao phase iv port), polar plot capability of a tug in indirect and direct mode of escort towing, use of association rules for cause-effects relationships analysis of collision accidents in the Yangtze River and VTS (Vessel Traffic Services) innovation, adaptation, and continued relevance.

The third chapter deals with new perspective for communication systems. The contents of the third chapter are partitioned into two subchapters. The third chapter deals with new perspective for communication systems. The contents of the third chapter are partitioned into two subchapters: a review of NAVDAT and VDES as upgrades of maritime communication systems; and detained vessels under Paris MoU: implementation of GMDSS.

The fourth chapter describes variety of challenges in safety of sea transportation: analysis of accidents during maritime transportation of cargo fumigated by phosphine, application of the BPMN models in maritime transport, mathematical model of hydrodynamic characteristics on the ship's hull for any drift angles, new passenger maritime transport system for Gulf of Cadiz, ship design challenges for ESPOMAR project (a review of available methods), study on applying numeric modelling CFD for fuel injection process of common rail system in marine diesel engine, the international convention for the control and management of ships' ballast water and sediments, a respectful way to tackle the problem of aquatic biodiversity in maritime transport, and theoretical research on mass exchange between an autonomous transport module and the environment in the process of transport from the seabed.

Each sub-chapter was reviewed at least by three independent reviewers. All were presented at the 13th International Conference TransNav 2019 on Marine Navigation and Safety of Sea Transportation, which held in Gdynia between 12 and 14 June 2019 .The Editors would like to express their gratitude to distinguished authors and reviewers of sub-chapters for their great contribution for expected success of the publication. They congratulates the authors and reviewers for their excellent work.

Advances in Marine Navigation

Analysis of Primary Position Validation in ECDIS System

D. Šakan, S. Žuškin, D. Brčić & S. Valčić
University of Rijeka, Rijeka, Croatia

ABSTRACT: The mandatory ECDIS transitional period has ended just recently. The technology matured and the progress towards the ECDIS as a primary means of navigation has continued. Safe and effective navigation mostly relies on accurate and valid positioning despite the system's technological advancement. In general, the frequency of plotting and monitoring of ship's position depends on two factors: employed navigational means and navigational areas. When using paper charts for primary means of navigation, position plotting frequency usually ranges from several minutes to one hour according to the Safety Management System (SMS) or other relevant requirements. In the case of ECDIS navigation, primary position is continuously displayed, mostly by the employment of satellite positioning means, with the possibility of secondary positioning display, depending on the source. Validation by position cross-checking is therefore essential for preventing overreliance on ECDIS primary position. In different navigational stages the navigator can use well known methods from celestial, coastal or electronic navigation. In the proposed paper, accuracy and complexity of positioning methods commonly used by the navigators are evaluated. The obtained results are correlated to primary positioning validation in ECDIS and compared with maritime user requirements, recommendations and respective resolutions. The research is accompanied with survey results on position checking methods conducted among the navigators. Results revealed that most of the navigators validate primary position, however there are misinterpretations between position validation, position fixing and secondary position source. Among stated, other overall and specific findings are presented and discussed. Future research directions are proposed in terms of the necessity of position checking, availability of positioning sources and further development of the system.

1 INTRODUCTION

Electronic Chart Display and Information System (ECDIS) implementation period for SOLAS vessels has ended on July 1st, 2018, marking the end of six years of transition. In many ways, ECDIS has changed the way how certain navigational tasks are done, but underlying principles for safe and efficient navigation are still the same. Improvements and benefits that ECDIS brought were accompanied with challenges, as with any developing and maturing technology. Opportunity to have a central focal point for assessing most of the information about the ship movement and surrounding environment is a major step forward for ever increasing complexity of present-day navigation.

One of the significant differences between the traditional paper chart and ECDIS navigation is the way position is presented. The continuous presence of position on the ECDIS display is a benefit compared to paper chart navigation. But this comes at a cost. The highly technical aspect of ECDIS technology can lull navigators to a false sense of accuracy for the presented positional and other information. To properly assess available information, nowadays navigators must retain traditional position fixing skills accompanied with in-depth knowledge about ECDIS performance. This is also reflected when analyzing validation of position in ECDIS. Currently, only Global Navigation Satellite System (GNSS) receiver is used for primary position source. To navigate safely, navigator must check this ever-present position with an independent position source or method. Additionally, secondary position source setting can be selected for comparison of primary position or as redundancy in case of failure. That is commonly a second GNSS receiver, but secondary position source is not limited only to GNSS, as is presented in later sections. To deal with this and other vessel operational challenges, companies had to develop and implement procedures and guidelines. These are part of the SMS which is mandatory under International Maritime Organization's (IMO),

International Safety Management ISM code (IMO, 1993). Between companies, SMS procedures can vary in scope and detail although they represent the same tasks and challenges.

So far, we have identified sources which can influence choice and interpretation of position fixing and validation from educational, regulatory and operational standpoint. Yet, none of the research regarding operational aspects of ECDIS isn't complete without feedback from the navigators and their opinions regarding position checking and validation.

In (Legić, 2016) difference of position cross-checking over position fixing was discussed in respect to international regulations and recommendations. In (Brčić et al., 2015) secondary positioning source, positioning methods and survey results were analyzed. Numerous papers are dealing with GNSS and other systems accuracy, but topic of ECDIS primary position and navigator methods for its validation is rarely elaborated. The aim of this paper is to provide overview on the accuracy of available and commonly used positioning methods by the navigators. This is correlated to primary positioning validation in ECDIS and analysis of survey answers from the end users. A brief overview of positioning methods is presented with observations on relevant SMS procedures from several shipping companies. It is followed by a survey results analysis section dealing with primary position validation, most common methods and correlation with secondary position source. In the discussion section, primary position validation is analyzed. Finally, findings are presented accompanied with proposals for future research.

2 NAVIGATION AND POSITION DETERMINATION

Navigation is a process of safe and efficient planning, execution and control of vessel movement. Depending on the navigational area in which ship sails, we can divide navigation into several distinctive phases: ocean, coastal and restricted waters phase (IALA, 2018). Restricted waters phase can be further subdivided into the harbour approach and port and inland waterways, which can be considered as different aspects of the restricted water phase. Ocean phase is usually beyond continental shelf with depth more than 200 m and distance from shore 50 or more nautical miles (NM). Coastal phase refers to areas which are less than 50 NM from shore or in the limit of the continental shelf with depth less than 200 m. Besides previously stated aspects, restricted water can also occur in coastal phase in straits.

Position fixing is a process where applicable navigational methods and techniques are used to obtain a position. Basically, position can be determined by lines of positions (LOP). However, position can also be calculated by dead reckoning or using inertial navigation systems.

When considering LOP methods, position is obtained by intersecting two or more LOPs, which can be either lines or curves. Howerer, at least three LOPs are required to obtain reasonably accurate position. Choice of navigational method and position fixing interval is based on the area in which ship sails. Generally, navigational methods can be defined as celestial or astronomical, terrestrial, electronic and dead reckoning or visual, radar, satellite, radio etc. Choice of the available method used results in different position accuracies. An overview of expected accuracies for position fixing techniques and systems is summarized in Table 1 which is originally presented in IALA Navigational manual (IALA, 2018). The values are derived from relevant IMO regulations or calculated based on 1 NM distance from the object.

Table 1. Fixing processes and systems (IALA, 2018)

Process	Typical accuracy (95% probability)	Accuracy at 1 NM (meters)
Magnetic compass bearing on light or landmark	±3° may deteriorate in high latitudes	93
Gyro-compass bearing on a light or landmark	0.75° X secant latitude (below 60° of latitude)	<62
Radio direction finder	±3° to ±10°	93 – 310
Radar bearing	±1° Stabilized presentation assumed. Reasonably steady craft.	32
Radar distance measurement	1 % of the maximum range of the scale in use or 30 meters, whichever is the greater	
LORAN-C / CHAYKA	Depends on conditions. Loran C was hyperbolic and provided 477m at the edge of coverage improving towards stations	
eLORAN	8 – 10 m differential Loran accuracies experienced at the port approach, typically available within 30 – 50km of a differential reference station	
GNSS	Generally, 3 – 5m for GPS	
DGNSS (ITU-R M.823/1)	1 – 3 m	
Dead Reckoning (DR)	Approximately one nautical mile for each hour of sailing	

Although several types of accuracies are defined in relevant resolutions and literature, for maritime applications absolute horizontal (geodetic or geographic) accuracy is commonly used. When considering radionavigation systems, accuracy requirements with 95% probability can be found in

IMO resolution A.1046(27) (IMO, 2011). Minimum maritime user requirements for general navigation and positioning with global navigation satellite systems (GNSS) and their respective absolute accuracies are presented in appendices of resolution A.915(22) (IMO, 2001).

Table 2. General navigation accuracy requirements

Navigation Phase	IMO resolution A.1046(27)	IMO resolution A.915(22)
Ocean	100 m	10 m
Coastal	10 m	10 m
Harbour approach	10 m	10 m
Harbour entrance	10 m	10 m
Port	N/A	1 m

2.1 *Overview of positioning methods*

The navigator should choose an appropriate method for positioning based on which navigational phase the vessels is in. For ocean navigation phase, besides satellite navigation systems, celestial navigation can be used. Basic instruments used are sextant, compass and chronometer alongside nautical almanac and navigational charts.

Technological advancement throughout late 20th and early 21st century significantly reduced usage of celestial navigation. Regardless of declining usage, there are still several advantages of celestial navigation. It can be used globally, it is not affected by electromagnetic interferences and denial or degradation of service. However, deliberate visual observation hampering on ships would be feasible by interested states. Today, on board ships software packages are commonly used for almanac data or celestial position calculation. Consequently, when used traditionally it can be considered not dependent on power.

Consequently, the possibility of taking celestial sights depends on weather conditions and visibility of sea horizon. Accuracy depends on the observer's skill, experience and precision of instruments used. Most commonly used method today is Marcq St. Hilaire intercept method where the position is derived from a graphical solution of celestial LOP intercepts. However, the position can also be calculated using direct methods such as Dozier method (Lušić, 2018). If possible, observations should be made once per watch or at least several times a day and during astronomical twilights. When it was used as a primary means of positioning in ocean navigation or as complementary to electronic position systems, considerable time and effort was spent on preparation, observation and evaluation of celestial fixes. An overview of pre-GPS celestial position fixing routines and challenges can be found in (Hohenkerk et al., 2012). Depending on the source, the accuracy of celestial fixes varies from ± 0,4 NM for experienced navigators (Hohenkerk et

al., 2012) up to ± 2 NM (Malkin, 2014). Commonly stated accuracy is around 1 NM (Brčić et al., 2015). Improved celestial methods are still developed (Pierros, 2018) despite declining usage. In (Lušić, 2018) author used ECDIS LOP function for plotting his proposed azimuth method without observed altitude of the celestial body even though this function is not intended for celestial observations.

Radar is used as a collision avoidance and navigation aid. To determine position several common methods are used. Those are LOPs by two or more bearings, two or more distances and by distance or bearing from conspicuous object. Bearings are measured by using Electronic Bearing Line (EBL). Distances are measured by using Variable Range Markers (VRM) or fixed range rings. Bearings and distances can also be obtained with the cursor. Accuracy according to IMO performance standards (IMO MSC.192(79), 2004) for ranges states that error should be less than 1.0% of range scale used or 30 m, whichever is greater. At 6 NM range that would result in single distance circle error value of approximately 110 m.

To obtain the most accurate range measurement, the radar must be properly set. Setup includes range and band selection, adjustment of gain and clutter values and pulse length selection. Also, the object must be properly identified, and distance properly measured. Additionally, object distance can be determined with echo referencing (Brčić et al., 2015) function. When considering radar bearings accuracy by before mentioned standards, the error can add up to 2.5°. Total bearing error is the sum of bearing accuracy with a value of ±1°, heading marker accuracy of ±1° and gyro input accuracy of ±0.5°. This would result in an error of around 480 meters at 6 NM range. Inherently, position by two or three radar bearings will be substantially less accurate than position obtained with radar ranges. Combined measurement of radar bearings and ranges is commonly measured from single conspicuous object. This method is quick and effective, but there is a possibility of wrong identification of objects and accuracy can considerably vary since the position is made on the intercept of single bearing and distance. Finally, there is also parallel indexing (PI) technique, a simple graphical method with parallel lines relative to ship course, which enables the navigator to quickly identify if the vessel is on the plotted route without the need to constantly determine position (Bole et al., 2014). This well-known and recommended non-positional method is especially useful in restricted waters with heavy traffic. Inherited advantage of radar is that distance measurement and relative bearing is independent on external sources (Brčić et al., 2015) but is affected by weather conditions, operator skill and experience and susceptible to intentional and unintentional interference.

For restricted waters phase, visual methods are also applicable. The position fix is commonly determined by taking bearings of conspicuous terrestrial objects. The position is determined as a point or an area by intersecting two or more LOPs. Preferably, three or more bearings should be taken to improve accuracy. Bearings can be taken by pelorus or bearing sight usually mounted on gyro compass repeater. If measured from ship's heading, bearings are termed, relative.

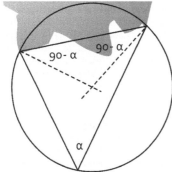

Figure 1. Horizontal angles

Besides bearings, horizontal angles between objects can be measured with bearing sight or a sextant. The resulting LOP is an arc of the distance circle from objects. Vertical angles between objects can also be measured with a sextant.

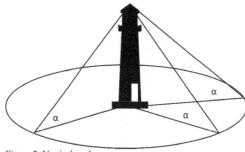

Figure 2. Vertical angles

There are still more visual methods available, both simultaneous or running fix, but they are applied rarely because of added complexity for measuring and graphical solutions (Lušić, 2013).

Considering the accuracy of visual observations, they are dependent on the skill of the observer, instrument errors and weather conditions. The most accurate LOPs would be obtained by bearing on leading line or similarly aligned transit bearing on two coastal objects. The navigator is limited to the visibility of objects and weather conditions.

Calculated accuracy of single gyro and magnetic bearings were presented in Table 1.

Dead Reckoning (DR) is a traditional method in which current position is determined from the last known position. It is calculated using the ship's speed, elapsed time from last known position not considering leeway and drift. The position is advanced by calculated value along the ship's steered course. Conversely, Estimated Position (EP) takes into consideration set and drift on ship movement and steering errors. Today, ECDIS systems have incorporated DR/EP function which uses speed log and (usually gyro) compass as input sources.

Satellite navigation is used in many different systems and areas of application, for positioning and very precise time measurement. The position is determined as an intercept of position spheres of at least three satellites. The radius of the sphere is range calculated as a function of time elapsed from the moment signal is sent from the satellite, and the known speed of signal propagation, until reception at the user's receiver. Global Navigation Satellite Systems (GNSS) is any satellite constellation that provides continuous Position, Navigation and Timing (PNT) capabilities. Satellite positioning is extremely precise when compared with previously described methods, however accuracy depends on satellite geometry, signal strength and obstructions, atmospheric conditions, end-user receiver design characteristics and quality. Also, as with any electronic position system, it is susceptible to outages, malfunctions, malpractice, degradation and intentional interferences. At present, USA's Global Position System (GPS) is the most used satellite navigation system. Despite terms GPS and GNSS are sometimes used interchangeably, GPS is not the only GNSS system available. Russian Globalnaya Navigatsionnaya Sputnikovaya Sistema (GLONASS) is also globally available. China's BeiDou Navigation Satellite System (BDS) started with provision of global PNT services on December 28th, 2018 – one year before the planned start. The European Union's Galileo system, with 22 Full Operational Capability (FOC) satellites is still in its initial service phase.

Additionally, positional accuracy can be enhanced by Ground Based Augmentation Services (GBAS) such as Differential GPS (DGPS), and Satellite Based Augmented services (SBAS) such as Wide Area Augmentation System (WAAS) and European Geostationary Navigation Overlay Service (EGNOS). DGPS accuracy from 1 to 3 meters can be achieved (Specht et al., 2018). The integrity of signals is determined with Receiver Autonomous Integrity Monitoring (RAIM) algorithm which uses additional signals available, besides those used for positioning.

Most of marine GNSS receivers are single frequency GPS receivers with DGPS capabilities (Glomsvoll and Bonenberg, 2017) (Januszewski, 2014), but less have SBAS capabilities. When we consider receiver capabilities at the present moment it is important to note that GNSS receivers on board ships are installed when the ship is built, meaning that these receivers are used for years. Despite the progress of Multi-GNSS signal availability and already defined shipborne multi-GNSS receivers (IMO MSC.115(73), 2000), it will take some time while most of the on-board equipment will benefit from expected improved accuracy (Dong et al., 2018). GNSS accuracy doesn't rely on navigator's skills or methods, however assessment of position quality does and can be incorrectly interpreted.

Currently, there isn't a reliable alternative to GNSS. There is ongoing process of development and implementation of a successor to mostly defunct terrestrial Long Range Navigation (LORAN-C) hyperbolic radionavigation system. Development of terrestrial high-power eLORAN system is part of PNT initiatives dealing with position and timing resilience. Positional accuracy for eLORAN is 10 m (Offermans et al., 2017) and even up to 5 m ("Research on alternative positioning navigation and timing in Europe," 2018). Benefits from the implementation of such backup system would resolve GNSS systems vulnerabilities and single source position reliance for contemporary applications.

2.2 *ECDIS position sources*

ECDIS is a navigation and information system which complies with relevant IMO resolutions (Žuškin et al., 2016). To be used as an alternative to paper charts for navigation it must comply with IMO's International Convention on Safety of Life at Sea (SOLAS) regulations V/19 and V/27. ECDIS displays primarily Electronic Navigational Charts (ENC) and other relevant information. Also, three mandatory devices must be connected. Firstly, an electronic position fixing system is required, currently a GNSS receiver. Secondly, a transmitting heading device, usually a gyro compass and finally, speed and distance measuring equipment, a speed log. With various options for data display and numerous mandatory and optional user settings, ECDIS is truly a complex system with many benefits for the navigators. Nonetheless, if not used with caution it can give a false sense of accuracy and lead to complacency. As such, considerable knowledge is required to reduce the possibility of misinterpretation of available data.

IMO resolution 232(82) states: "ship's position should be derived from a continuous positioning system of accuracy consistent with the requirements of safe navigation. Whenever possible, a second independent positioning source, preferably of a different type, should be provided. In such cases, ECDIS should be capable of identifying discrepancies between the two sources" (IMO MSC.232(82), 2006).

Commonly, two GNSS (preferably DGPS) receivers are used for the primary and secondary source. Although independent, they are based on the same satellite technology. An alarm will be raised when the deviation limit is exceeded between the two sources regardless of the cause. Nonetheless, there is always a possibility that some system-wide or local interference could influence both sources. If not validated by an alternative independent positional method this could pass undetectable. If the primary position fails, the system should automatically select available secondary source or switch it to DR, raising the alarm.

Furthermore, position and time should be marked when ECDIS switched to dead reckoning. Besides primary, secondary and DR and EP positions, there is a possibility of graphical methods plotted by various plotting functions available in ECDIS using methods described in the previous section. Additionally, the position can be validated by using the ECDIS Radar Image Overlay (RIO), if available. Any discrepancies should reveal positional error if the RIO image is not correlated with corresponding shore features.

2.3 *Primary position validation*

While ECDIS positional sensors are working properly, positional methods are used for cross-checking, not position fixing, since the position is constantly updated and presented on the ECDIS display (Legieć, 2016). With that notion, it is even more important that primary position sensor is cross-checked and validated. As it was explained in the previous paragraph, there are several ways that cross-checking can be accomplished.

To further explore how navigators perceive position fixing and checking, we have evaluated SMS procedures from one of the major container liner companies, large crude oil tanker and LNG carrier companies, chemical tanker and bulk carrier company. Although describing the same procedures, there are variations in scope and detail regarding position fixing intervals and methods. Also, some procedures explicitly state the difference between position checking and fixing and describe how the navigator should validate positions. Prescribed methods include comparison of primary with secondary position, position fixing methods, depth monitoring, track history, visual confirmation and sensor integrity monitoring. There are even procedures and drills in case of positional and total ECDIS failure, which is especially important for paperless vessels. It is also worth to mention that the container liner company has a policy of an updated folio of port approach charts on paperless vessels. Most detailed and descriptive SMS procedures regarding position checking were by container liner and chemical tanker companies. In container liner SMS procedures there is an actual statement that position fixes: "should be limited to important events only (watch handover, entry and exit of buoyed channels, course alterations, MOB etc.)" and that position fixing system should be checked at least once per watch. In chemical tanker company SMS procedures there is a statement that position fixing checks should be carried out from few to several minutes. For the ocean phase (termed deep sea in that SMS procedures) GPS position should be verified on both ECDIS' every hour. In coastal waters position should be verified by visual and radar means, approximately every twenty minutes. These are maximum position check intervals which could be shorter, and they reflect common intervals from position fixing methods. When evaluating LNG carrier company SMS, detailed position fixing intervals and methods are described. For significant geographical straits and passages position fixing interval should not exceed 10 or 15 min, just as for approach/departure to berth and pilot station.

3 ECDIS EHO SURVEY RESULTS

The survey from which elaborated questions are analyzed, started in 2014 for attendees of ECDIS Generic Course (MC IMO 1.27 2010) at the Faculty of Maritime Studies in Rijeka. The survey was aimed primarily towards assignment of attendees into ECDIS simulator working groups, based on their experience, knowledge and familiarization level. Afterwards, the questionnaire was also provided to respondents on various international shipping companies, making the survey international. Coming from various experience backgrounds, shipping companies and ship types, respondent profiles correspond to the diversity of possible targeted ECDIS navigators. Survey results and findings have resulted in several research papers considering various aspects of ECDIS usage, education and training (Žuškin et al., 2016, Brčić et al., 2016, Brčić and Žuškin, 2018).

3.1 Questionnaire and question overview

The questionnaire is entitled: "ECDIS Survey Analyses: Experience, Handling, Opinion" or abbreviated "ECDIS EHO". It is comprised of 26 questions covering: profile-defining – introductory questions, operational questions and questions regarding system handling and knowledge. Also, respondents share their experiences and opinions.

Responses regarding position validation were collected in the period 2015 – 2018.

3.2 Respondents profile

Respondents who completed the survey have different maritime operational and ECDIS experience backgrounds. For further profiling, they were classified by their rank, work and ECDIS experience. Out of 181 respondents who completed the survey there were 55 Masters (M), 36 Chief officers (CO), 25 Second officers (2O), 5 Third Officers (3O), 20 Deck Cadets (DC), 3 Environmental officers (EO), 14 Port State Control Officers (PSCO), 1 Pilot (P), 1 Cargo Officer (CGO), 1 Safety Officer (SFO), 1 Maritime Safety Consultant (MSC) and 19 Unidentified respondents (UR). Undefined respondents come from all previously stated ranks.

To adequately distribute respondents to the target group several excluding stages were conducted based on the following criteria.

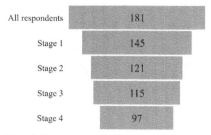

Figure 3. Target group selection stages

In the first stage, respondents without ECDIS operational experience were excluded. In the second stage, to obtain the most relevant answers regarding position validation during navigation, only Masters, 2nd Officers, 3rd Officers and Deck Cadets were selected, thus excluding other ranks and positions. In the third stage, respondents with missing profile data such as working experience were further excluded. In the fourth stage, respondents whose job duties do not include working with ECDIS, do not have navigational duties or aren't forming part of the navigational watch were also excluded. Finally, the target group was reduced to 97 respondents who satisfied set exclusion criteria.

Table 3. Target group rank distribution

Master	Ch. Off.	2nd Off.	3rd Off.	Cadet	Total
38	22	19	4	14	97

Furthermore, the respondents were grouped to three major subgroups: Masters, Officers (OOWs) and Cadets.

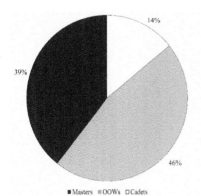

Figure 4. Target subgroup distribution

From the target group respondents' answers following two questions were evaluated:
1. By which means you perform the checking of position displayed on the ENC? (Q1)
2. Do you use the setting secondary positioning source on your ECDIS system? If YES, which device/system is it? (Q2)

Although the primary aim is the evaluation of Q1, it is important that we also correlate position checking with the usage of secondary position sources stated in Q2.

As expected most respondents answered affirmative regarding ECDIS position checking. Only a few respondents answered negatively as it is reflected in the respected percentage. The percentage of respondents who did not provide an answer is labelled with abbreviation N/A.

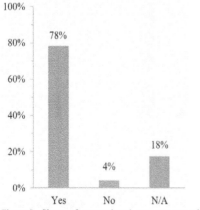

Figure 5. Share of respondents' answers regarding ECDIS position checking (Q1)

When analyzing subgroups, the percentage of affirmative answers decreases when looking from Masters' subgroup to Cadets' subgroup, however is still mostly affirmative. Masters subgroup has the highest percentage of affirmative answers and

lowest percentage of respondents who did not answer to the proposed question. When evaluating negative responses by rank subgroup, only the OOWs group had negative answers. Four respondents who answered No were all Chief Officers. In 2015 and 2016, when they completed the survey their age was from 55 to 67 years with working experience ranging from 25 to 40 years. Additionally, they also provided negative answers to Q2.

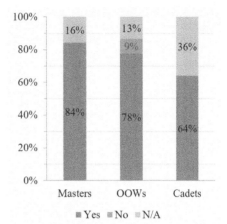

Figure 6. Subgroup distribution shares on ECDIS primary position checking (Q1)

The majority of respondents are using radar for checking position presented on ENC in ECDIS, followed by visual methods and GPS. Also, many respondents stated using multiple methods for position checking. Thus, the percentage of the represented method corresponds to cumulative share from all respondents who gave an affirmative answer (Yes) to Q1 for each method. Still, the results represent the method's occurrence in the target group.

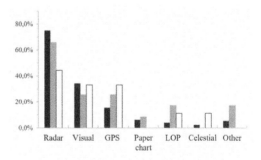

Figure 7. Primary position checking methods share for each subgroup

Masters subgroup has the highest usage of radar position checking methods, with OOWs following closely. The Deck Cadets have the equal distribution when considering most frequent methods, but we must restate that 36% did not provide any answer regarding methods and their subgroup had the lowest number of respondents (14). Beside three major groups of methods: radar, visual and satellite (GPS) some respondents answered with lines of position (LOP). Respondents did not provide further explanation, so to avoid ambiguity which methods are used, LOP answers were grouped as a separate category. Some respondents have paper charts, either as one of the possible options in conformance with SOLAS regulation regarding navigational equipment or as an addition to ECDIS. They stated checking of ECDIS position with plotting on paper charts. Finally, from all respondents in the target group, only one Master and Deck Cadet have answered that they check position by celestial methods. The remaining Other category includes answers such as Parallel Index lines technique (PI), sounding, manual position fixes, Estimated Position or survey positioning program. Since there were mostly single answers except PI and manual fixing, they were grouped in category *Other*.

When analyzing the usage of secondary position source setting Q2 affirmative answers are considerably lower than checking primary position. Also, there is a much larger percentage of negative answers while there is a similar percentage of respondents who did not respond compared to Q1.

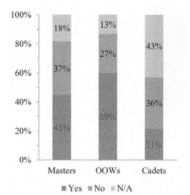

Figure 9. Subgroup distribution shares on ECDIS secondary position (Q2)

The highest percentage of respondents answering affirmatively is in OOWs subgroup. Deck Cadets have the largest number of unanswered and negative answers.

When considering the selection of secondary position sources, from respondents who answered affirmatively that they use secondary source, the results are obtained as follows. Masters have predominantly answered GPS as a secondary source, but somewhat of 30% gave multiple answers regarding secondary position source besides GPS. Around 50% of OOWs also use GPS with several answers regarding multiple methods. Deck Cadets have highest share value for Dead Reckoning, but the share represents only two respondents.

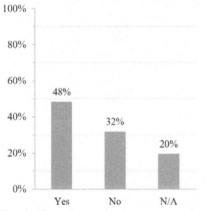

Figure 8. Share of respondents' answers regarding ECDIS secondary position source setting (Q2)

Less than 50% of Masters' use secondary position setting with a slight difference of just 8% higher than negative answers.

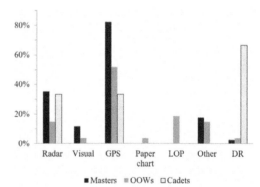

Figure 10. Secondary position sources selection for each subgroup

Also, respondents from all subgroups stated multiple methods for secondary position source with very few answers stating Dead Reckoning. Finally, an interesting observation is that none of the respondents stated GNSS as an answer when questioned about satellite navigation systems; all respective answers were termed as GPS.

4 DISCUSSION

So far, we have presented background information regarding available position methods, their advantages, accuracies and survey results. The results confirmed expectations that most of the navigators are doing primary position checking. This is, of course, a result of good seamanship and company policies set by relevant IMO regulations. Inexperience might be a source of insecurity regarding statements about position checking for future officers (Deck Cadets). Masters and OOWs, as more experienced navigators predominantly stated that they check primary position. On average 40% of all respondents from the target group have stated that they use multiple methods for checking the primary position. Again, the lowest percentage of multiple methods was in the Cadet group.

When considering methods, only two answers are stating celestial observations and positions. This can be attributed to several factors: lack of skills, the complexity of methods, dependence on favorable position fixing conditions and large positional error. Although celestial navigation is still taught and is part of certification exams, unfortunately, it is seldom used. Celestial observations are stated explicitly as part of position fixing methods or as a secondary means of fixing for the ocean (deep sea or open sea) phase of navigation in three out of five previously mentioned SMS procedures. It is interesting why the results of the survey did not show larger usage since these are the only methods besides DR that are available as a secondary positioning fixing method. Although the use of celestial position methods is declining, progress for using ("Electronic Sextant G-Stell,") contemporary technology to augment traditional navigation instruments (Bobkiewicz, 2008) should not stop.

The second observation is radar methods usage for primary position checking. For Masters and OOWs the ratio of radar methods used is almost twice as much when compared to visual methods. This could indicate that navigators either choose radar because of presumed accuracy and practicality due to Radar Image Overlay (RIO) in ECDIS, availability even in low visibility or by object identification with electronic chart overlay on some radars.

Analysis of secondary position source usage showed that 48% of target group respondents are using secondary position setting while 32% do not and 20% did not provide an answer. When correlating this distribution with SMS procedures previously described in chapter about primary position validation, secondary positioning fixing methods are mentioned explicitly, but ECDIS secondary positioning source is not. Also, primary and secondary means of navigation are defined either as second ECDIS or paper chart folio. In chemical tanker company SMS procedures, it is stated that vessels which use ECDIS as a primary navigation system and have two types approved ECDIS systems "shall have them both permanently connected to one or two Global Positioning Systems (GPS)" so although indirectly, we can assume that this implies usage of GPS for secondary position source. The option "one or two" leads to possible doubt of secondary position source selection. From the same SMS procedures, there is a paragraph dealing with vessels which use paper charts as a primary navigation system, which explicitly states primary and secondary position fixing methods.

In previously examined SMS procedures from all five evaluated companies, only container liner company SMS procedures state that "Unless in fault, the GPS/DGPS position is shown continuously. Otherwise, ECDIS switches to dead reckoning". We can argue and imply that navigators have relevant knowledge from education and training and that there is no need to prescribe every step or possibility in the navigational process, such as the difference between position fixing, cross-checking, validation and checking of integrity. Nonetheless, procedures deal with other details of navigation systems: position methods, fixing intervals and contingency plans in case of primary means failure. However they mostly exclude details regarding the primary and secondary position, cross-checking and validation and give very general directions.

The survey results show that most navigators who use secondary positioning setting choose GPS as a secondary position source, but it seems that there is some misinterpretation what that term implies. When we analyze answers from respondents who answered Yes to proposed Q2, some 18% stated usage of other sources besides 2nd (D)GPS or DR. Many respondents gave answers about radar fixing and overlay, visual methods etc. Those answers are describing position cross-checking and not secondary position source setting. If we correlate this result to 52% of respondents which did not provide answers, we must observe that for many respondents, there is no clear distinction from validating the primary position and choosing secondary positioning source.

Finally, we must discuss the validation of primary position when dealing with positional accuracy. Navigators are familiarized about inherent limitations of electronic navigation systems from literature, relevant regulations and recommendations (Maritime and Coastguard Agency, 2008) and accordingly by company procedures. Yet, as we presented in previous chapters, the so called traditional methods have much lower accuracies and depend on many factors, including navigator's competence, for optimizing that accuracy. Typical accuracies of (D)GNSS and of other methods at 1 NM distance from the object were presented in

Table 1. When we compare the accuracies, it is obvious that under normal conditions position from radar or visual LOPs will be far less accurate than (D)GNSS. If we are aware of possible inaccuracies and errors, can we say that we checked primary GNSS position? This is the reason why it is very important to assess how navigators interpret accuracy of presented GNSS position both in ECDIS and in dedicated GNSS receivers.

5 CONCLUSION

In proposed paper the contemporary sources on accuracy were evaluated for available positioning methods. The results were correlated with relevant regulations and shipping companies' procedures on primary position validation. Finally, survey responses on primary position validation and secondary positioning source from the ECDIS navigators were analyzed. Results confirmed that most navigators check primary position and that radar methods are most frequently used, followed by visual methods and GNSS. It is important to note that celestial methods are rarely used. Conversely, only about half of the navigators use secondary position setting. Besides GPS, respondents stated usage of multiple secondary positioning sources such as positions from radar and visual methods. This indicates that there is no clear distinction between secondary position source, secondary position fix and validation among respondents. This is also reflected in SMS procedures. Only one out of five companies' surveyed SMS procedures has explicitly stated secondary position source.

Primary position validation is very important for safe and efficient navigation. The difference between position fixing and position checking is not purely semantic. This also applies to secondary positioning source and secondary fixing. Additionally, there must be clear understanding how accurate is a position from different positioning methods, especially in terms of near-future advances in satellite navigation. ECDIS on-board procedures must reflect technological changes and can't be just a simple upgrade from paper chart navigation.

Observations from the presented survey and analysis direct to several areas of future research. This research will be directed towards evaluating current and near-future progress and implementation of viable independent positioning system besides GNSS, such as eLORAN. Also, possibilities for new or augmented methods for celestial and terrestrial navigation using contemporary technology will be explored. Furthermore, primary and secondary position questions will be expanded in future surveys. A better understanding of navigators' interpretation of procedures, knowledge of positional accuracy, advantages and limitations of

systems and validation methods should result in better and optimized procedures. Finally, the ECDIS system should adapt to navigators' needs with greater possibilities for position accuracy interpretation and validation, especially with near-future developments of GNSS systems.

ACKNOWLEDGEMENTS

This study has been financially supported by the University of Rijeka under the Faculty of Maritime Studies projects. Authors are grateful to all the officers of the navigational watch for their time and willingness for the fulfillment of the survey, discussions and their opinions. Authors believe that their answers have an immense significance for the appropriateness of the research deliverables.

REFERENCES

Bole, A.G., Wall, A., Norris, A., 2014. Radar and ARPA manual: radar, AIS and target tracking for marine radar users, 3rd edition. ed. Butterworth-Heinemann, Oxford.

Brčić, D., Kos, S., Žuškin, S., 2016. Partial structural analysis of the ECDIS EHO research: The handling part. Presented at the 24th International Symposium on Electronics in Transport (ISEP).

Brčić D., Kos S., Žuškin S.: Navigation with ECDIS: Choosing the Proper Secondary Positioning Source. TransNav, the International Journal on Marine Navigation and Safety of Sea Transportation, Vol. 9, No. 3, doi:10.12716/1001.09.03.03, pp. 317-326, 2015

Brčić, D., Žuškin, S., 2018. Towards Paperless Vessels: A Master's Perspective. Annals of Maritime Studies / Pomorski Zbornik 55, 183–199.

Bobkiewicz P.: Estimation of Altitude Accuracy of Punctual Celestial Bodies Measured with Help of Digital Still Camera. TransNav, the International Journal on Marine Navigation and Safety of Sea Transportation, Vol. 2, No. 3, pp. 279-284, 2008

Dong, Z., Cai, C., Santerre, R., Kuang, C., 2018. An Enhanced Multi-GNSS Navigation Algorithm by Utilising a Priori Inter-System Biases. The Journal of Navigation 71, 339–351. https://doi.org/10.1017/S0373463317000637

Electronic Sextant G-Stell, STARNAV. URL https://www.starnav.fr/site/en/electronic-sextant-g-stell/ (accessed 2.12.19).

Glomsvoll, O., Bonenberg, L.K., 2017. GNSS Jamming Resilience for Close to Shore Navigation in the Northern Sea. The Journal of Navigation 70, 33–48. https://doi.org/10.1017/S0373463316000473

Hohenkerk, C., Kemp, J., Nibbs, B., 2012. Astro Navigation Remembered. The Journal of Navigation 65, 381–395. https://doi.org/10.1017/S0373463312000033

International Association of Marine Aids to Navigation And Lighthouse Authorities 2018. NAVGUIDE 2018 Marine Aids to Navigation Manual, 8th ed. IALA

International Maritime Organization, 2001. Resolution A915(22), Revised maritime policy and requirements for a future global navigation satellite system (GNSS): London IMO

International Maritime Organization, 2000. Resolution MSC.115(73) Adoption of the revised performance standards for shipborne combined GPS/GLONASS receiver equipment: London IMO

International Maritime Organization, 1993. International management code for the safe operation of ships and for pollution prevention (International Safety Management (ISM) code): London IMO

International Maritime Organization, 2011. A.1046(27) Worldwide radionavigation system: London IMO

International Maritime Organization, 2004. Resolution MSC.192(79) Adoption of the revised performance standards for radar equipment. London IMO

International Maritime Organization, 2006. Resolution MSC 232(82) Adoption of the revised performance standards for Electronic Chart Display and Information systems (ECDIS). London IMO

International Maritime Organization, 2010. Model Course 1.27, Operational use of Electronic Chart Display and Information systems (ECDIS). London IMO

Januszewski, J., 2014. Shipborne satellite navigation systems receivers, exploitation remarks. Scientific Journals of The Maritime University of Szczecin, Zeszyty Naukowe Akademii Morskiej w Szczecinie 112, 67.

Legień, W., 2016. Position Cross-Checking on ECDIS in View of International Regulations Requirements and OCIMF Recommendations. TransNav, the International Journal on Marine Navigation and Safety of Sea Transportation 10.

Lušić, Z., 2013. The Use of Horizontal and Vertical Angles in Terrestrial Navigation. Transactions on Maritime Science, 2(1), pp. 5 - 14. doi: 10.7225/toms.v02.n01.001.

Lušić, Z., 2018. Astronomical position without observed altitude of the celestial body. Journal of Navigation 71, 454–466. https://doi.org/10.1017/S037346331700073X

Malkin, R., 2014. Understanding the Accuracy of Astro Navigation. The Journal of Navigation 67, 63–81. https://doi.org/10.1017/S0373463313000520

Maritime and Coastguard Agency, 2008. MGN 379 (M+F) Navigation: Use of Electronic Navigation Aids.

Offermans, G., Bartlett, S., Schue, C., 2017. Providing a Resilient Timing and UTC Service Using eLoran in the United States. Navigation, Journal of the Institute of Navigation 64, 339–349. https://doi.org/10.1002/navi.197

Pierros, F., 2018. Stand-alone Celestial Navigation Positioning Method. Journal of Navigation 71, 1344–1362. https://doi.org/10.1017/S0373463318000401

Research on alternative positioning navigation and timing in Europe, 2018. 2018 Integrated Communications, Navigation, Surveillance Conference (ICNS), Integrated Communications, Navigation, Surveillance Conference (ICNS),2018.https://doi.org/10.1109/ICNSURV.2018.8384 887

Specht, C., Pawelski, J., Smolarek, L., Specht, M., Dabrowski, P., 2018. Assessment of the Positioning Accuracy of DGPS and EGNOS Systems in the Bay of Gdansk using Maritime Dynamic Measurements. The Journal of Navigation 1–13. https://doi.org/10.1017/S0373463318000838

Žuškin, S., Brčić, D., Kos, S., 2016. Partial structural analysis of the ECDIS EHO research: The safety contour. Presented at the 7th International Conference on Maritime Transport: Technological, Innovation and Research Maritime Transport '17.

Comparison between Simulator Modelled and Empirical Ship Squat Prediction

M. Baric & L. Grbic
University of Zadar, Zadar, Croatia

R. Mohovic & D. Mohovic
University of Rijeka, Rijeka, Croatia

ABSTRACT: One of the most important component in prediction of ship dynamic draught is squat. Ship squat is steady hull downward displacement due to relative water movement around ship hull. That phenomena causes different pressure distribution around the hull and causes water depressions in which ship sinks. Prediction of the ship squat can be made using proposed empirical formulas or navigational simulators. Empirical formulas are based on physical model testing and field measurements for certain fairway and ship configurations, while navigational simulators use complex mathematical and hydrodynamic modelling for given scenario to calculate and predict ship squat. The accuracy of the simulator models are tested on ship models and have proven accuracy in predicting ship squat. This paper analyses differences between modelled and calculated ship squat for two general types of vessels, in a narrow channel and for a different ships speed.

1 INTRODUCTION

Ship squat is a steady downward displacement consisting of a translation and rotation due to the flow of water past the moving hull. This water motion induces a relative velocity between the ship and the surrounding water that causes a water level depression in which the ship sinks. Shallow water and channel banks significantly increase these effects. The velocity field produces a hydrodynamic pressure change along the ship that is similar to the Bernoulli effect since kinetic and potential energy must be in balance [1]. This means, that ship squat is composed of decrement of ship under keel clearance (UKC) and trim change.

That change, generally, occurs at one part of the ship. It is rule of the thumb, that maximum squat occurs at the bow for full form ships and at stern for fine formed ships. However, initial trim of the vessel may also have the influence where the maximum squat occurs [2].

Estimation of ship squat depends which method is used to calculate squat and also parameters of the ship and fairway. The most important ship parameter is speed through the water. It is generally estimated that ship squat is increased with the square of the ship speed, so if the ship speed is increased by two, the squat will increase by four times. Other important ship parameters are draft and block coefficient.

The fairway parameters that influence ship squat are depth and width. The rule of the thumb states that squat is significant if the ship draft and water depth ratio is less than 1.5. That means, the ship will squat in deeper water but will not be affected due to small risk of touching the bottom. For easier calculation, fairway configuration is idealised into three types [3]. First type is unrestricted fairway, which represents larger water area with small depts. and not restricted by width (generally width greater than eight times the ship´s beam). Second type is restricted channel, with smaller depth and submerged banks. Third type is canal, with restricted depth and width (usually inland access between two larger water bodies).

2 EMPIRICAL SQUAT CALCULATION

Most of the empirical formulas are based on model testing and real ship measurements in different fairway configurations. In 2014. PIANC [3] recommended seven empirical methods to predict ship squat. Only two methods (Authors Barrass and Romisch) predict ship squat at bow and stern and other five (Authors HUUska/Guliev, ICORELS; Eryuzulu and Yoshimura) predict only ship squat at the bow. In this paper those two methods, which predict ship squat at the bow and stern, will be used in comparison.

Author Barrass [4] proposed following formula for calculation maximum squat at bow and stern:

$$S_{max} = \frac{Cb \cdot V^2}{100/K}, \qquad (1)$$

where:
Cb – block coefficient,
V – ship speed (in knots),
K – dimensionless coefficient

Dimensionless coefficient K represents value for predicting fairway parameters and is calculated using following formulae:

$$K = 5,74 \cdot S^{0.76}, \qquad (2)$$

where:
S – fairway blockage factor.

The fairway blockage factor represents proportion of cross sectional area of a fairway occupied by the vessel. A value of S=0.1 is equivalent for a wide fairway, and S=0.25 is a value for a restricted fairway. Where the maximum squat occurs, in Barrass formulae depend on ship block coefficient (Cb). If the block coefficient is less than 0.7 ship will squat with stern, if the block coefficient is greater than 0.7 ship will squat with bow.

Barrass also presented formulas for determining the value of squat on the other ship end (side opposite of maximum squat), formulae for determining of amount of squat due to mean body sinkage and formulae for determining the amount of squat due to ship dynamic trim.

Squat on other end is calculated using following formulae:

$$S_{other\,end} = K_{other\,end} \cdot S_{max}, \qquad (3)$$

where the *other end* represents coefficient defined as:

$$K_{other\,end} = \left[1 - 40\left(0.7 - Cb\right)^2\right] \qquad (4)$$

The amount of body sinkage is calculated using following formulae:

$$S_{mean} = K_{mean} \cdot S_{max}, \qquad (5)$$

where the K_{mean} represent coefficient defined as:

$$K_{mean} = \left[1 - 20\left(0.7 - Cb\right)^2\right] \qquad (6)$$

The amount of ship squat due to the ship dynamic trim is calculated:

$$S_{dynamic\,trim} = K_{dynamic\,trim} \cdot S_{max}, \qquad (7)$$

where the $K_{dynamic\,trim}$ represent coefficient defined as:

$$K_{mean} = 40\left(0.7 - Cb\right)^2 \qquad (6)$$

Author Romisch [5] [6] developed formulae for calculating ship squat at the ship bow and stern. Data was gatherer using physical models for all three types of idealised fairway types.

Given squat formulas at the ship bow and stern are:

$$S_{bow} = C_V \cdot C_F \cdot K_{\Delta T} \cdot T, \qquad (7)$$

$$S_{bow} = C_V \cdot K_{\Delta T} \cdot T, \qquad (8)$$

where:
C_V – correction factor for ship speed,
C_F – correction factor for ship shape,
$K_{\Delta T}$ – correction factor for squat at ship critical speed,
T – ship draft.

Correction factor can be calculated using following expressions:

$$C_V = 8\left(\frac{V}{Vcr}\right)^2\left[\left(\frac{V}{Vcr} - 0.5\right)^4 + 0.0625\right], \qquad (10)$$

$$C_F = \left(\frac{10Cb}{Lpp/B}\right)^2, \qquad (11)$$

$$K_{\Delta T} = 0.155\sqrt{h/T}, \qquad (12)$$

where:
V – ship speed (m/s),
Vcr – critical ship speed (m/s),
Lpp – ship length between perpendiculars,
B – ship breadth,
h – water depth.

In this method where the maximum squat will occur, depends on value of C_F. If the value of C_F is equal to 1 the maximum squat will occur at stern, and if greater than 1 at ship bow.

If we compare these two methods, it is noticeable that Barrass method is more user friendly and mostly used for determining squat. That's why this method will be compared with simulator modelled ship squat.

3 SIMULATOR MODELLED SHIP SQUAT

Nowadays navigational simulator is a tool which offer most realistic result. The reason for that is because the models are based on data gathered on physical models or real ship data. In this paper Transas NaviSailor 5000 simulator was used to gather ship squat data.

First part of simulator testing is choosing ship models and fairway type. For this research six types of ship models were chosen, three full form vessels (Cb over 0.7) and three fine form vessels (Cb under 0.7). In Table 1 is complete data for used ship model.

Table 1. Ship models used in research

Ship model name	D (t)	LOA (m)	B (m)	T (m)	Cb
Oil tanker	77100	242.8	32.2	12.5	0.82
VLCC2	321260	332	58	20.82	0.81
VLCC4	137092	249.9	44	15.4	0.83
Cont4	132540	346.98	42.8	14	0.65
Cont5	86900	299	37.1	13	0.62
CarCarrier4	24186	196.4	31.1	7.1	0.55

Fairway is prepared using Transas model package "Model Wizard". This allows user to construct desired fairway and to use it on simulator. For this purpose fairway was unrestricted, with constant depth. The reason is to generally test empirical squat formulae and its accuracy.

Second step is preparing exercise and conditions. Environmental conditions were not included and were disabled in order to gather ship draft change due to squat. In simulations three different ships speeds were used, 5 knots, 7.5 knots and 10 knots. Ship under keel clearance (UKC) was simulated with two different depths. Due to different ship models and different drafts, the UKC was expressed through water depth/ ship draft ratio (h/T). Water depth, for each model, was set to ratio h/t=1.2 and h/t=1.5. It is rule of the thumb that water ratio over h/t=1.5 represent "deeper" water and under "shallower" water.

Third and the last step is performing the simulations. For the each simulation UKC at the bow and stern was measured and recorded. In total for each model there were six different simulation settings which were performed.

4 RESULTS

As it was stated in previous paragraph in all simulation, as a result, UKC at the ship bow and stern was measured. In Table 2 and table 3 result are shown.

Table 2. UKC results for all ship models and depth draft ratio h/T=1.2

Model name	Depth draft ratio $h/T = 1,2$				
	Ship speed (kn)	UKC bow (m)	UKC stern (m)	Max squat (m)	Squat according to Barrass (m)
Oil tanker	5,35	2,29	2,19	0,52	0,47
	7,45	2,18	2,06	0,76	0,91
	10,10	1,88	1,82	1,30	1,67
VLCC2	5,40	3,96	3,98	0,39	0,31
	7,40	3,87	3,91	0,55	0,59
	9,80	3,72	3,81	0,80	1,04
VLCC4	5,70	2,86	2,91	0,39	0,36
	7,40	2,80	2,86	0,50	0,61
	9,90	2,56	2,76	0,84	1,08
Cont4	5,90	2,73	2,63	0,24	0,30
	7,30	2,72	2,58	0,30	0,35
	9,72	2,66	2,49	0,45	0,61
Cont5	7,40	2,52	2,45	0,23	0,34
	9,80	2,47	2,38	0,35	0,60
CarCarrier4	5,13	1,44	1,38	0,14	0,14
	7,82	1,35	1,30	0,31	0,34
	10,00	1,22	1,18	0,56	0,55

Table 3. UKC results for all ship models and depth draft ratio h/T=1.5

Model name	Depth draft ratio $h/T = 1,5$				
	Shop speed (kn)	UKC bow (m)	UKC stern (m)	Max. squat (m)	Squat according to Barrass (m)
Oil tanker	5,40	6,14	6,04	0,32	0,24
	7,40	6,07	5,95	0,48	0,45
	9,85	5,91	5,81	0,78	0,80
VLCC2	5,40	10,20	10,21	0,41	0,24
	7,40	10,14	10,15	0,53	0,44
	9,80	10,05	10,08	0,69	0,78
VLCC4	5,70	7,51	7,55	0,34	0,27
	7,40	7,45	7,50	0,45	0,45
	9,90	7,32	7,43	0,65	0,81
Cont4	5,90	6,95	6,86	0,19	0,23
	7,32	6,93	6,82	0,25	0,35
	9,75	6,90	6,75	0,35	0,62
Cont5	7,40	6,44	6,37	0,19	0,34
	9,90	6,40	6,33	0,27	0,61
CarCarrier4	5,13	3,65	3,60	0,15	0,14
	7,90	3,60	3,55	0,25	0,34
	10,1	3,52	3,48	0,40	0,56

As it is noticeable in results, maximum squat occurs at the bow for the vessels with the block coefficient over 0,.7 (*Cb>0,7*) and at the stern for the vessel with the block coefficient less than 0.7 (*Cb<0,7*).

Next step is to compare the squat results gathered with simulator to those calculated using Barrass empirical formulae. The result are shown in Figure 1 and in Figure 2.

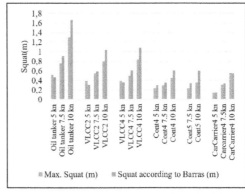

Figure 1. Comparison of maximum squat from simulation and calculated squat using Barrass empirical formula for depth draft ratio h/T=1.2

The results in Figure 1. show that the biggest difference occurs at greater speed. The reason for that may be the safety facor included into the empirical formula. Also, empirical formulae are general and may not be the same for different types of ship, mainly due to different hydrodinamic forces forming around the ship hull. For the example, ship model CarCarrier4 has the same amount of squat in simulator as calculated using empirical formula.

In Figure 2. where the depth draft ratio is h/T=1.5 it can be seen that also at larger speed there is bigger difference between simulator modelled squat and empirically calculated squat. Also, amount of simulator modelled and empirically calculated squat for ship model Oil tanker is the almost same.

This confirms that empirically modelled squat is quite general, and the squat will differ for various fairway conditions (as per UKC) and hull shape.

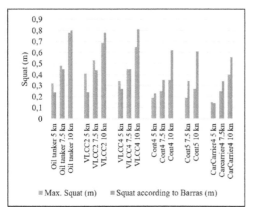

Figure 2. Comparison of maximum squat from simulation and calculated squat using Barrass empirical formula for depth draft ratio h/T=1.5

5 CONCLUSION

Ship squat is well known phenomena which is over the years well explained and can be calculated quite accurately. Due to that empirical methods are quite accurate and provide quick and reliable tool for determining ship squat. Navigational simulator provide the latest tool for predicting ship dynamical movement due to included proved and tested mathematical and hydrodynamic models. That is confirmed trough this research, which proved the reliability of analysed method. That enables the end user, either on ship or land, to determine ship squat and to enable adequate level of safety of navigation. The future work will include testing additional ship models and other deterministic methods for calculating ship squat.

REFERENCES

[1] *Hydrodynamic Problems of Ships in Restricted Waters | Annual Review of Fluid Mechanics*. Available at https://www.annualreviews.org/doi/10.1146/annurev.fl.10. 010178.000341.
[2] C.B. Barrass, *Thirty-two years of research into ship squat*, Squat-Workshop: Aspekte der Unterkielfreiheit in Analyse und Anwendung, Elsflether Schriften zur Seeverkehrs-und Hafenwirtschaft, Bd 3 .
[3] *Harbour Approach Channels Design Guidelines*. Report of Marcom Working Group 49, 2014.
[4] C.B. Barrass, *Ship Squat and Interaction*, Withersby Publishing & Seamanship, Livingstone, UK, 2009.
[5] K. Romisch, Section I - Inland Navigation, 1977.
[6] K. ROMISCH, *Der Squat in bergrenzten Fahrwasser-Betrachtung aus hydromechanischer Sicht*, Heft 10 (1993), pp. 58–62.

Environmental Impact of New Maritime Routes Cadiz-Huelva-Algarve: Preliminary Study

D. Marín, E. Nebot, L.A. Fdez-Güelfo, D. Sales & A. Querol
University of Cadiz, Cadiz, Spain

ABSTRACT: In the present paper it is studied the environmental impact of connecting the region of Cadiz, Huelva and the Portuguese Algarve's coast by means of maritime transport. For this purpose, it is applied a bottom-up methodology to find out and compare the distance and times of the land and maritime routes, the fuel consumption and the emission of 23 polluting compounds for a car-type and 20 types of boats. Additionally, taking the emission levels into consideration, the Air Quality Index (AQI) is calculated and a new concept of an "Environmental Index (EI)" that condensates them is formulated and presented. Taking Cadiz as the home origin, the greatest differences in the emissions occur in the boats that have a lowest consumption and in those maritime routes whose distances are smaller than the ones of the current existing land routes.

1 INTRODUCTION

The transportation of products and passengers has produced an economic and social development without precedent all over the world (European Union 2011). However, current and future trends in transport show that its development should be considered as unsustainable because it presents different forms of impact on both the environment and people's health (World Health Organization 2000).

Road transport is the preferred mean of transport for Spanish citizens (Spanish Statistical Office 2009) due to the benefits offered against public transport (more comfort and less travel time). As a result, this makes motor vehicle traffic the main source of pollution in urban areas (World Health Organization 2000). However, shipping had been considered less environmentally damaging than other means of transport. Although, when the ships navigate have the capacity to cause significant pollution in coast (Eyring et al. 2005) and, in some cases, due to breezes, that transport contaminants far inland, affecting both areas at the same time (Eyring et al. 2009). There is no consensus on which form of transport is more environmentally sustainable, and some scientist claim that maritime transport emissions are the main source of pollution in port cities (Eyring et al. 2009) and others believe that road transport emissions are more dangerous (Viana et al. 2014).

The most important compounds emitted by road transport and shipping are generated in the combustion motors and can be grouped by their impacts: O_3 precursors (CO, NO_x, NMVOC, CH_4), greenhouse gases (CO_2, CH_4, N_2O, BC), acidifying gases (SO_2, NO_x, NH_3), particulate matter and heavy metals (among them, Hg, Cd, As, Cu) (European Environment Agency 2016).

In view of the above, mobility should move towards modes of transportation with less externalities, more capacity to move passengers and become a real alternative to traditional modes of transportation.

To sum up, in the present paper the Environmental Impact Study (EIS) of this European project is carried out. Thus, the aim of the study is to determinate the Environmental Impact associated to a new maritime route for passenger transport connecting Cadiz, Huelva and Algarve. In order to achieve this: 1) The impacts of maritime and road transport are identified and compared between them; 2) The influence of fuel consumption and the routes on the emissions are evaluated; 3) A ideal boat having a low environmental impact is chosen.

Figure 1. General and local ubication of study area with all routes and ports

2 METHODS AND MATERIALS

Figure 1 shows the study area. Nowadays the population of the region exceeds one million (Institute of statistics and cartography of Andalusia 2017, Statistics Portugal 2011) and also the tourism sector has a particular importance, being the most touristic regions of their respective countries (Eurostat 2017). Both, inhabitants and tourists would be potential customers of this service.

2.1 Distance and time calculation of routing

Differentiating between the two means of transport:
- Private vehicle: the distance is known throught the Michelin Guide (2018), which prioritizes the shortest and fastest routes linking all towns. Travel times are calculated on maximum permitted speed on each section of road.
- Boat: the maritime routes between towns are obtained with the Google Earth Pro searching for the shortest distances. 20 real ferries were selected from the Journal *Significant Small Ships* (Royal Institution of Naval Architects, RINA), showed in Table 1. So, travel times are obtained with the speed of each ferry and the distance of the shortest routes.

2.2 Estimation of fuel consumption and emissions

The Bottom-up methodology used by Eyring et al. (2009) is employed to obtain the emissions of 23 polluting compounds. This methodology calculates the emissions according to the distances and emissions factors for each contaminant.

The compounds studied, appearing in parenthesis the reference where the emission factors have been extracted from, are:
- Private vehicle: CO, SO_2, N_2O, NMVOCs, CH_4, NH_3, PM, dioxins, furans, As, Cd, Cr, Cu, Hg, Ni, Pb, Se, Zn (SEI 2014); BC (Ježek, et al. 2015);

CO_2 (Institute for Energy Diversification and Saving 2017).
- Boat: BC (International Council on Clean Transport 2014); NH_3; dioxins and furans (Institute for Energy Diversification and Saving 2017); HCB, PCB, OC, CO, SO_2, N_2O, NMVOCs, CH_4, PM, As, Cd, Cr, Cu, Hg, Ni, Pb, Se, Zn, CO_2 (Dalsoren, et al. 2009).

Table 1. Main characteristics of the 20 real ferries.

Ferry	Passengers	Speed km/h	Fuel consumption L/h
Horizon 9	238	49	240
Farasan	650	59	2612
Taha Dua and Taha satu	64	46	210
Jacque Cartier	202	54	549
Osman Gazi-1	1200	69	6300
Sea Scape 1	300	33	190
The Grand Canal Shoppes	411	78	2276
Gavea I	900	33	250
Aremiti 5	700	65	2069
Lake Express	248	63	2069
Bocayna Express	450	63	2837
First travel XXXI/XXXII	354	30	400
Ramon Llull	462	63	3428
Salten and Steigtind	214	61	1135
Zephyr	464	56	851
New York	400	65	650
The cat	900	74	5827
Euroferrys Pacifica	951	69	6265
The Lynx	900	70	6856
Highspeed 2	620	74	3546

Data is extracted from RINA magazines (from SSS2000 until SSS2014).

As a result, the estimation of fuel consumption and emissions are calculated as follows:
- Private vehicle: The Spanish automotive market is divided unequally between gasoline and diesel cars, with a specific fuel consumption (L/km) for each kind of vehicle. According to the European Automobile Manufacturer Association Report (2016) in Spain diesel cars entail 56.9 % of all the

cars, while gasoline cars entail 43.1 %, consuming 0.0625 and 0.0769 L/km, respectively. For these reasons, all calculations in this study are made for a vehicle-type, which combines the characteristics of diesel and gasoline cars in function of the diesel/gasoline rate. The results of both the fuel consumption and the emissions take into account the real distribution of the Spanish automotive market.

After all, the fuel consumption ($FC_{d/g}$), in L, of diesel or gasoline cars is calculated with the follow equation:

$$FC_{d/g} = D * fc_{d/g}. \tag{1}$$

where D = distance between the municipalities, in km; d = diesel vehicles; g = gasoline vehicles; and fc = the specific fuel consumption for vehicles, in L/km.

Next, the fuel consumption of diesel or gasoline cars is known, the fuel consumption (FC), in L, of the ideal car is obtained by the following equation:

$$FC = FC_d * P_d + FC_g * P_g. \tag{2}$$

where P = rate of each type of vehicles on the market.

The emission factors are needed to determine the emissions of pollutant gases. All the emissions factors employed have been pondered taking into account the ratio of diesel and gasoline vehicles. Thus, the emissions for a specific compound, for any route ($Q_{i,j}$), in their pertinent unit for mass is determined using the next equation:

$$Q_{i,j} = FE_i * D_j. \tag{3}$$

where D_j = the distance of each trip, in km; and FE_i = is the emission factor for a specific compound.

- Boat: for each of the 20 selecterd ferry, the specific fuel consumption or SFC (L/h) is given by the RINAs database. So, Total Fuel Consumption, in L, for each route ($TFC_{j,k}$) is calculated applying the next equation:

$$TFC_{j,k} = \frac{D_j}{V_k} * SFC_k. \tag{4}$$

where D_j = the distance of each boat trip, in km; V_k = the specific speed of each ferry, km/h.

For each of the 23 compounds, the emissions for a specific compound, for any route and any boat ($Q_{i,j,k}$), in their pertinent unit for mass is obtained using the follow equation:

$$Q_{i,j,k} = FE_i * TFC_{j,k} \tag{5}$$

2.3 Classification of the compound by their main impacts

Following the literature, the 23 contaminants are classified in the next groups: CO2-eq, SO2-eq, Equivalent Heavy Metals, PM, NOx and AQI.

"CO2-eq" is used to unify the gases that produce greenhouse effect as CO_2, CH_4, N_2O and BC. The weighting factors that these gases receive are respectively: 1, 25, 298, 900 (Intergovernmental Panel on Climate Change 2007). That means that 1 gr of methane is equivalent or has the same impact on greenhouse effect to emission of 25 gr of CO_2.

"SO2-eq" is used to unify the gases that contribute to water acidification as SO_2, NO_x y NH_3. The weighting factors that these gases receive are respectively: 1, 0.7, 1.88 (European Commission, 2006). The concept of "Equivalent Heavy Metals" or HM-eq (Hakanson 1980) employs toxic effects factors for the following heavy metals: Hg, Cd, As, Cu, Pb, Cr and Zn. The value of the factors that these compounds receive are respectively: 40, 30, 10, 5, 5, 2, 1. The term "AQI" or Air Quality Index (Environmental Protection Agency 2014) let us know if the air that is breathed implies some risk to human health. The more AQI value, the more harmful results for human. Normally, this index is calculated with immission values of 5 compounds (SO_2, NO_2, PM, O_3 and CO). Furthermore, European Union has established air quality standards that each compound concentration should not exceed to protect human health. Using that standards and emissions values a sub-index for each compound is obtained. Air Quality Index takes the value of the highest sub-index. All information is condensed in the next formula (Department of Environment, Territory and Infrastructure of Galicia 2014):

$$AQI = Max\left(0.286 * Q_{SO2}; 0.67 * Q_{PM10}; 10 * Q_{CO}; 0.5 * Q_{NO2}; 0.556 * Q_{O3}\right) \tag{6}$$

where Max = the maximum value reached by the 5 compounds.

However, AQI has been modified in this work because immision values were replaced by emission values, obtained with equation 3 for vehicles and equation 5 for boats. But there are not emission factors neither for O_3 nor NO_2, these are omitted. As a result, AQI is calculated with formula:

$$AQI = Max\left(0,286 * Q_{SO2}; 0.67 * Q_{PM}; 10 * Q_{CO}\right) \tag{7}$$

2.4 Environmental Index

An Environmental Index (EI) is presented with the objective of unification those six indexes in a single value, being easier to data analyze. In this new index weighting factors are assigned to each index based on literature. The new index is obtained with the next formula:

$$EI = CO_2 eq\, S * 4 + SO_2 eq\, S * 2 + HMeq\, S * 1.75 + AQI\, S * 2.25) \quad (8)$$

where S = Standard Component of EI.

The standardization for each index is achieved dividing their values by the maximum occupancy of mode of transport concerned (private car or one of twenty boats). Subsequently, it is divided by the maximum value reached on any route, and any mean of transport. Taking CO_2-eq as example, for the route to Tavira and for *Gavea I*, that would be:

$$CO_2 eq\, S = \frac{Q_{CO_2 eq, Gavea I, Tavira}}{\left(900 * Max\, Q_{CO_2, j, k}\right)}$$

2.5 Environmental comparison between maritime and land transport

It is taken into consideration:
- The influence of fuel consumption on emissions is studied.
- The influence of routes distance on emissions is evaluated.

In both, emissions and fuel consumption must be expresses per passenger transported (by car or by ship).
- The average vehicle occupancy in Europe was found about 2 persons per vehicle (European environment Agency 2003).
- It is calculated the minimum demand of passengers that is needed to transport by ship for reaching the same emissions associated with the transport of passengers in private car (Paravantis et al. 2008).

3 RESULTS AND DISCUSSION

All the maritime and land routes, for each means of transport (20 boats and the private car) and emissions for each compound from a list of 23, have been calculated for the nine municipalities considered (Fig. 1). However, in the present paper only Cadiz´s data are presented. Cádiz is the most populated city among the options considered and its tourist sector is the most relevant, with more than 18,000 accommodations (including hotels, hostels and shelters). All of them could be potential users of the maritime transport service.

3.1 Time and distances

Considering the private transport, Cadiz –as origin of all routes- has the lowest values in the accumulated time and the accumulated distances, being 1439.50 km and 16.63 h, followed by Conil (1481 km and 16.98 h) and Chipiona (1577.50 km and 18.08 h).

In addition, as it can be seen in Figure 2, fifteen out of twenty of these maritime routes can be traveled from Cadiz and the rest of villages faster than the private transport, that means in less than 2.08 h. The *Grand Canal Shoppes*, *Highspeed 2* and *The Cat* are the fastest ships. Maritime routes mean 58.79 % of the total distance of road routes from Cadiz.

3.2 Gaseous and particulate pollutants

The five less polluting means of transport obtained with each group are, ranked from the lowest to the highest, as follows:
- CO_2-eq emissions (kg/pax): *Gavea I* (19.84), *Sea Scape 1* (45.23), *Horizon 9* (48.91), *New York* (59.68) and Private transport (71.61).
- SO_2-eq emissions (kg/pax): Private transport (0.13), *Gavea I* (0.57), *Sea Scape 1* (1.29), *Horizon 9* (1.40) and *New York* (1.70).
- PM emissions (kg/pax): Private transport (0.01), *Gavea I* (0.04), *Sea Scape 1* (0.10), *Horizon 9* (0.11) and *New York* (0.13).
- NO_x emissions (kg/pax): Private transport (0.17), *Gavea I* (0.37), *Sea Scape 1* (0.82), *Horizon 9* (0.89) and *New York* (1.08).
- HM-eq emissions (g/pax): *Gavea I* (0.05), *Private transport* (0.06), *Sea Scape 1* (0.11), *Horizon 9* (0.12) and *New York* (0.14).
- AQI (pax^{-1}): *Gavea I* (0.43), *Sea Scape 1* (0.98), *Horizon 9* (1.06), *New York* (1.29), and *Zephyr* (1.68).

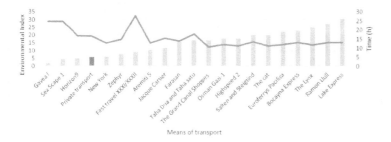

Figure 2. Total amount of Environmental Indexes (vertical column) and total amount of time (continuous line), in hours, for each means of transportation, taking Cádiz as origin.

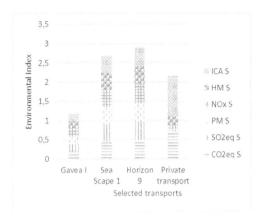

Figure 3. Environmental Index for 4 means of transport (3 ships and a car-type), for the summa of all the routes.

To jointly analyze this data the Environmental Index (EI) is employed. According with EI values (Eq. 8), *Gavea I, Sea Scape 1* and *Horizon 9* are the only ships of RINAs database whose affections on the environment is lower than the private transport's affections. The value of EI for these ships means 34.43, 78.50 and 84.90 % of the reached value by the car (Fig. 2).

These results are related to fuel consumption and the number of passengers that can be transported. The fuel consumption is lower in *Sea Scape 1* (190 L/h) than *Gavea I* (250 L/h), but the last one can move more passengers. This fact causes that EI value *for Gavea I* is smaller than the value of Sea Scape 1. As result, *Gavea I, Sea Scape 1* and *Horizon 9* have been chosen in order to carry out the Environmental Impact Study.

Also, the values of the standard components or sub-indexes (for example, CO_2-eq S) of the main compounds that are discharged to the environment by the ships and the private transport are showed in Figure 3. On the basis of their environmental impact and taking Cadiz as home port, the means of transport can be classified, in terms of their EI values, from least to greatest: *Gavea I* (1.18), *Private transport* (2.16), *Sea Scape 1* (2.68) *and Horizon 9* (2.90). It is clear that ships and cars are completely different. The ships show a constant proportion among the indexes, their values being greater as the fuel consumption per passenger increases. On the other hand, in the case of the private transport, the AQI and CO_2-eq are the most relevant indexes.

3.3 *Environmental comparison between maritime and land transport*

Comparing the private transport with the three selected boats, it is revealed that there is a strong, direct correlation between the average fuel consumption (L/pax) and the average emissions (kg/pax, of CO_2-eq, SO_2-eq, PM, NO_x and HM-eq). Therefore, this fact justifies the result shown in Figure 2, where *Gavea I, Sea Scape 1* and *Horizon 9* appear as the least pollutant ones.

In addition, it is determined that the greater the difference between the distances of maritime and land routes are, the more difference for emissions will be.

The study of the minimum demand of passengers from which the ship is able to equal or reduce the emissions from the private transport is focused on routes from Cadiz to Barbate, Mazagon, Tavira and Faro (Fig. 3). This study concludes that only for CO_2-eq, HM-eq and NO_x it has been possible to find a passage for which the emissions balance is favorable for the ships. The case of NO_x is special, since only *Gavea I* is able to reduce the emissions from the private transport (Fig. 4). In that Figure, negative values mean reduction in the NO_x released to the atmosphere.

The route that needs less passengers to avoid excess of emissions to the atmosphere, in comparison with car, is the route of Mazagon, with 646 passengers. However, Tavira and Faro need 819 and 861 passengers, respectively. There are no occupancy levels that avoid emissions for the route to Barbate.

Obviously, at high occupancy levels, the emissions per passenger of the different compounds to the atmosphere are lower. For this reason, it is necessary that this service becomes attractive for the citizens so that they decide to switch from private cars to the ferry.

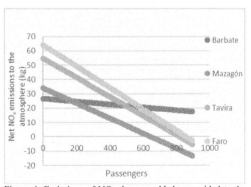

Figure 4. Emissions of NO_x that are added or avoided to the atmosphere at various occupancy levels for Gavea I (Cadiz as origin).

4 CONCLUSSIONS

Recalling that it is a preliminary study, the main conclusions obtained are:

- Cadiz has been chosen as home port for the studied routes. This is because is the most populous city in the region, the tourist sector is notable and geographically is the best located. The last one originates that Cadiz has the lowest values in the accumulated time and the accumulated distances in both maritime and land routes.
- An Environmental Index (EI) has been defined from four standard sub-indexes (CO_2-eq S, HM-eq S, AQI S, SO_2-eq S) which show that *Gavea I, Sea Scape 1* and *Horizon 9* have better EI than the car. Their values for EI suppose 34.43, 78.50 and 84.90 % of the reached value by the private transport.
- The better choice is *Gavea I* as it has the lowest value for the EI, the lowest contribution to climate change on terms of CO_2-eq, the lowest input of heavy metals and the lowest affections to the air quality of all the studied ships. Additionally, it manages to offset the emissions of NO_x for an occupation higher than 819 passengers.
- Maritime transport between Cadiz and the studied destinations in Huelva and the Portuguese Algarve´s region is environmentally viable. The ship that would make these routes should have similar characteristics to *Gavea I* and should have a tourist flow that ensures a high occupancy. Therefore, this mode of transport should be accompanied by support and promotion of public institutions and private tourist promoters.

ACKNOWLEDGEMENT

The authors would like to thank the European Regional Development Fund (ERDF) of the European Union (EU) for co-financing the project "ESPOmar" under the "INTERREG POCTEP 2014-2020 program". We also like to thank to all researchers and staff of the different public organizations involved on the project for their dedication and professionalism.

REFERENCES

Dalsøren, S.B. Eide, M.S. Endresen, Ø. Mjelde, A. Gravir, G. & Isaksen, I.S.A. 2009. Update on emissions and environmental impacts from the international fleet of ships: the contribution from major ship types and ports. *Atmospheric Chemistry and Physics* 9: 2171-2194.

Department of Environment, Territory and Infrastructure of Galicia. 2014. *Instrucción Técnica del Cálculo del Índice de Calidad del Aire.*

Environmental Protection Agency. 2014. *Air Quality Index. A guide to Air Quality and Your Health.*

European Automobile Manufacturers Association. 2016. *Economic and Market Report. EU Automotive Industry.*

European Commission. 2006. *Annex 5. Environmental impacts analysed ad characterization factors.*

European Environment Agency. 2003. *TERM 2002 29 EU-Occupancy rates of passenger vehicles.*

European Environment Agency. 2016. *EMEP/EEA air pollutant emission inventory guidebook.*

European Union. 2011. White Paper on Transport.

Eurostat. 2017. *Eurostat regional yearbook*

Eyring, V. Isaksen, I.S.A. Berntsen, T. Collins, W.J. Corbett, J.J. Endresen, O. Grainger, R.G. Moldanova, J. Schlager, H & Stevenson, D.S. 2009. Transport impacts on atmosphere and climate:Shipping. *Atmospheric Environment* 44 (37): 1-37.

Eyring, V. Köhler, H.W. van Aardenne, J. & Lauer, A. 2005. Emissions from international shipping:1. The last 50 years. *Journal of Geophysical Research* 110 (D17): 1-12.

Häkanson, L. 1980. An Ecological Risk Index for Aquatic Pollution Control: A Sedimentological Approach. *Water Research* 14: 975-1001

Institute of statistics and cartography of Andalusia. 2017.

Institute for Energy Diversification and Saving. 2017. *Guía de Vehículos Turismo de venta en España, con indicación de consumo y emisiones de CO_2.*

Intergovernmental Panel on Climate Change. 2007. *Climate Change 2007. The physical science basis.*

International Council on Clean Transport. 2014. *The state of clean transport policy. A 2014 synthesis of vehicle and fuel policy developmnets.*

Ježek, J. Katrašnik, T. Westerdahl, D. & Mocnik, G. 2005. Black carbon, particle number concentration and nitrogen oxide emission factors or random in-use vehicles measeured with the road chasing method. *Atmospheric Chemistry and Physics* 15: 11011-11026.

Michelin Guide. 2018. https://www.viamichelin.es/

Paravantis, J. Sambracos, E. & Ntanon, S. 2008. *Energy Consumption and Carbon Dioxide Emisions of a Suburban Coastal Transport System. AMunich Personal RePEc Archive.*

Royal Institution of Naval Architects. 2018. *Journals about Significant Small ships.*

Sistema Español de Inventario de Emisiones. 2014. *Inventarios Nacionales de Emisiones a la Atmósfera 1990-2012.*

Spanish Statistical Office. 2009. *Encuesta de Hogares y Medio Ambiente.*

Statistics Portugal. 2011.

University of Cadiz. 2011. https://espomar.uca.es/proyecto/descripcion-general/

Viana, M. Hammingh, P. Colette, A. Querol, X. Degraeuwe, B. de Vlieger, I. & van Aardenne, J. 2014. Impact of maritime transport emissions on coastal air quality in Europe. *Atmospheric Environment* 90: 96-105.

World Health Organization. 2000. *Transport environment and health.*

World Health Organization. 2011. *Burden of disease from environmental noise. Quantification of healthy life years lost in Europe.*

Optimal Route Planning Algorithm for Coastal Ships Considering Ocean Environment and Grounding

W. Lee, W. Yoo, S.H. Ham & T. Kim
Seoul National University, Seoul, Republic of Korea

ABSTRACT: To find the safe and economical route of coastal ships, not only fuel oil consumption, but also the risk of grounding should be taken into consideration. This paper proposed an optimization method for the most economical route planning while avoiding grounding near coast. A modified A* algorithm was utilized to search the route in real space considering grounding risk. Under keel clearance (UKC) method was utilized to avoid grounding, and this method was investigated using the ocean depth information from electronic navigational chart. An optimization method with sequential quadratic programming was performed in order to compute the efficiency of ship operation by applying ISO 15016 with available weather information such as wind, wave and current. This paper computed the optimal route considering the ocean environment and grounding between two specific ports. As a results, the efficiency and the safety were compared between the proposed route and the past route.

1 INTRODUCTION

Marine accidents can cause loss of humans and properties, and affect the marine environment adversely. It has been reported that 85% of grounding and collision accidents occur due to human error, and 22% of grounding accident occurs due to the deficient route plan (Tsou, 2010).

In the past, navigators determined the route plan considering paper charts related to the ocean environment. However, the manual route plan has weakness in perspective of safety and efficiency. First, the way to consider grounding is different for each navigator. The manual route plan has a dependence on navigator's experience. An inexperienced navigator possibly causes a grounding accident due to a dangerous route plan. Second, the manual route plan does not consider ocean environments such as wave, wind, and current. Even though warning or caution for weather is used to make a route plan, ocean environment for efficiency is not considered properly.

For these reasons, the optimization of the route plan is essential to enhance navigational safety and efficiency. A typical way to optimize the route plan was to minimize fuel oil consumption using A* algorithm (Jung & Rhyu, 1999). Hinnenthal & Clauss (2010) used the Pareto optimization method to optimize the route considering fuel oil consumption and safety. Lin et al. (2013) proposed 3D dynamic programming for determining the

economical route by minimizing fuel consumption. Szlapczynska (2015) set the fuel consumption and the ship stability as the objective function and utilized evolutionary algorithm by dividing the voyage interval. Lee et al. (2018) utilized a genetic algorithm that simultaneously optimized the route and speed by modifying a great circle route.

Since these studies have been performed based on weather routing, the proposed algorithms of other researches are not possible to the route plan in the coastal sea. Unlike weather routing that navigates for dozens of days, coastal routing navigates in a few hours. The computation time is important due to the short voyage time in coastal routing.

Therefore, this study formulated a two-stage optimal route planning algorithm in order to attain short computation time. At the first stage, the route is optimized considering dangerous zones such as lands and shallow depth areas in electronic navigational charts. At the second stage, the optimum engine rpm is computed considering ocean information to reduce fuel oil consumption. Thus, optimal route planning algorithm searches the safe route while minimizing fuel oil consumption in a short time.

2 OPTIMAL ROUTE PLANNING ALGORITHM

2.1 *Problem description*

The factors considered in route planning for coastal ships are grounding, ocean environment and regulations.

First, grounding is considered since shallow areas and lands exist in coastal seas. The grounding criteria are considered with the size of the ship and the under keel clearance(UKC). UKC is the margin between seafloor and bottom of the ship.

Second, the ocean environment is considered to optimize the route planning using ship resistance and propulsion performance. The ship usually navigates along the shortest route to reduce fuel oil consumption since the fuel oil consumption increases as the voyage distance increases. However, if the ocean environment of the shortest route is bad, the fuel oil consumption of the shortest route will be more than that of the route bypassing to the bad ocean environment. Therefore, the fuel oil consumption of the bypass route is less than that of the shortest route even though the length of the bypass route is longer than that of the shortest route.

Finally, the regulations for the coastal ships is considered in order to design an optimal route planning algorithm similar to the actual route planning in the coast. These regulations include the speed limitation around ports, the area for the fishery, and the area with the shipwreck.

2.2 *Route search*

In order to optimize the route plan in a short time, this research proposed a two-stage optimal route planning algorithm. At the first stage, the A* algorithm is utilized as a route search algorithm. The A* algorithm is a widely used grid-based algorithm since it has a short computation time when finding a local optimum close to the global optimum. In this research, the modified A* algorithm was utilized to search the route in real space since electronic char information is a polygon in real space.

The A* algorithm computes the evaluation function at all position that can move from the current position. Then, the current position is moved to a position with the lowest evaluation function. The A* algorithm repeats computation of the evaluation function and movement to the point with lowest evaluation value until the ship arrives at the destination. The evaluation function is expressed as the sum of the cost function and the heuristic function. The cost function is the fuel oil consumption from the departure to the next position. The heuristic function is the fuel oil consumption from the next position to the destination. The braking power of the heuristic function is computed by using the assumption that the ship's speed and

ocean environment are the same until the destination. The cost function and heuristic function is expressed as Equation 1 and 2:

$$g(\mathrm{x}) = f_{rate} P_B(n) \frac{d_{\mathrm{x}}}{v(n)} \tag{1}$$

$$h(\mathrm{x}) = f_{rate} P_B(n) \frac{d_{\mathrm{x-final}}}{v(n)} + risk(\mathrm{x}) \tag{2}$$

where x is next position, f_{rate} is fuel oil consumption rate, P_B is brake power, n is engine rpm, d_{x} is the distance between departure and next position, $d_{\mathrm{x-final}}$ is the distance between next position to the destination, v is the ship speed, and $risk$ is grounding risk.

The function $risk$ is considered to decide grounding risk using UKC. UKC is set by multiplying the ship's maximum draft by a constant coefficient. If the UKC of the position is insufficient, the value of $risk$ is infinity so that the move to the position is prohibited.

2.3 *Fuel oil minimization*

After the route is determined, the optimum engine rpm should be found to minimize the fuel oil consumption. Since the speed and resistance of the ship are changed when the ship meets the waves, winds, and currents, it is necessary to calculate accurate additional resistance. However, the calculation of the additional resistance is so complicated and takes much time. Therefore, in this study, the total resistance in calm water, the wave additional resistance, and the wind additional resistance are considered to estimate the power. The fuel oil consumption is estimated using power considering wave, wind, and current. The wave additional resistance and the wind additional resistance is computed based on ISO 15016. ISO 15016 is a standard for reducing a gap between the real test and the model test (ISO, 2002).

The propulsion performance is estimated by the additional resistance based on ISO 15016 and the current speed. In order to navigate at the desired speed of the ship, the brake power is essential for the estimation of fuel oil consumption. The brake power is computed using the ship's model test and additional resistance based on ISO 15016. The current is used to adjust the ship's speed by utilizing the relative speed.

3 CASE STUDY

The simulation was performed to navigate along routes from Mokpo to Jeju using an optimal route planning algorithm. In this simulation, the fuel oil consumption of the optimal route was compared with that of the past route. In the scenario, the test ship departs from Mokpo at 10:00 and arrives at Jeju at 14:30, and the voyage time is 4 hours and 30 minutes. According to the coastal ship regulations, the ship's speed would not exceed 12 knots near Mokpo and Jeju port. Fig. 1 shows the optimal route computed by the route search algorithm and the past route. The optimal and past routes were marked in yellow and red.

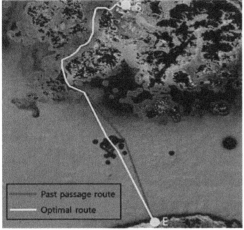

Figure 1. Past route (red) and optimal route (yellow)

Figure 1 shows that the past route and the optimal route were different from each other. There is little difference between the optimal route and the past route around the departure point since there are islands and lands. There is a big difference between the optimal route and the past route in the area with no island. In the above simulation, three scenarios are compromised to apply the fuel oil consumption minimization. The fixed engine rpm was applied to the past route in scenario 1. The fixed engine rpm was applied to the optimal route in scenario 2. The optimized the engine rpm was applied to optimal route in scenario 3.

Table 1. The comparison of each scenario.

Setting	Fuel oil consumption		Estimated
	ton	percentage	time of Arrival
Scenario 1	16.62	100%	4.49(hour)
Scenario 2	16.09	96.77%	4.49(hour)
Scenario 3	15.99	96.18%	4.48(hour)

Table 1 shows the results of the optimal route planning algorithm for the three scenarios. The fuel oil consumption ratio was set based on scenario 1 with the fixed engine rpm in past route. In comparison with the fuel oil consumption of Scenario 1, it was confirmed that the fuel oil consumption of scenario 2 was reduced by 3.23%

and that of Scenario 3 by 3.82%. In comparison with the fuel oil consumption of scenario 1 and scenario 2, the fuel oil consumption was reduced by navigating to the optimal route rather than navigating to the past route. In Scenario 2, the length of the route was shorter than that of Scenario 1 so that the engine rpm was lower and the fuel consumption was decreased. The fuel oil consumption in scenario 3 is smaller than that in scenario 2 due to the optimized engine rpm.

4 CONCLUSION

The two-stage optimal route planning algorithm was formulated in order to search the route without grounding and minimize fuel oil consumption in a short time. At the first stage, A* algorithm was utilized to compute the optimal route. At the second stage, the fuel optimization was utilized to minimize fuel oil consumption. In the fuel optimization, the objective function is the fuel consumption, the design variable is the engine rpm, and the constraint is the estimated time of arrival and the coastal ship regulation. It was confirmed that the optimized route has no grounding point and that the fuel oil consumption is reduced by 3.23% when compared with the past route. It is confirmed that the fuel oil consumption is reduced by 0.59% due to rpm optimization. It was found that the fuel oil consumption was reduced by 0.63 ton and the fuel ratio by 3.82% when the route search algorithm and the fuel minimization were applied.

In this research, we found that the route is optimized to avoid grounding and minimize fuel oil consumption in a short time. However, this research only considers grounding among various risks such as grounding, collision and stability problem. Future research will be needed to optimize route considering the traffic conditions around the route and the stability of the ship.

REFERENCES

Galor, W. 2011. Determination of Dynamic Under Keel Clearance of Maneuvering Ship. *The Journal of Konbin.* 8(1): 53-60.
Hinnenthal, J. & Clauss, G. 2010. Robust Pareto Optimum Routing of Ships Utilising Deterministic and Ensemble Weather Forecasts. *Ships and Offshore Structures*, 5: 105-114.
Hayashi, S. & Kuwajima, S. 1991. A stranding Avoidance System Using Radar Image Matching – Development and Experiment. *The Journal of Navigation.* 44: 205-212.
ISO. 2002. Guidelines for the assessment of speed and power performance by analysis of speed trial data. *ISO/DIS 15016*: 1-45.
Jung, J. S. & Rhyu, K. S. 1999. A study on the Optimum Navigation Route Safety Assessment System Using Real

Time Weather Forecasting. *Journal of the Korean Institute of Office Automation* 4(1): 38-46.

Lee, S., Roh, M., Kim, K., Jung, H. & Park, J. 2018. Method for a Simultaneous Determination of the Path and the Speed for Ship Route Planning Problems. *Ocean Engineering*, 157: 301-312.

Lin, Y. H., Fang, M. C. & Thong, S. K. 2013. The Optimization of Ship Weather-routing Algorithm Based on the Composite Influence of Multi-dynamic Elements. *Applied Ocean Research*, 43: 184-194.

Rho, M. 2013. Determination of an Economical Shipping Route considering the Effects of Sea State for Lower Fuel Consumption. *International Journal of Naval Architecture and Ocean Engineering.* 5(2): 246-262.

Shao, W., Zhou, P. & Thong, S. 2012. Development of a Novel forward Dynamic Programming Method for Weather Routing. *Journal of Marine Science and Technology.* 17(2): 239-251.

Szlapczynska, J. 2015. Multi-objective Weather Routing with Customised Criteria and Constraints. *Journal of Navigation.* 68(2): 338-354.

Tsou, M. 2010. Integration of a Geographic Information System and Evolutionary Computation for Automatic Routing in Coastal Navigation. *Journal of Navigation.* 63(2): 323-341.

Park, J. & Kim, N. 2014. A Comparison and Analysis of Ship Optimal Routing Scenarios considering Ocean Environment. *Journal of the Society of Naval Architects of Korea.* 51(2): 96-106.

Psaraftis, H. N. & Kontovas, C. A. 2014. Ship Speed Optimization: Concepts, Models and Combined Speed-routing Scenarios. *Transportation Research Part C: Emerging Technologies.* 44: 52-69.

Probabilistic Modeling of Ship-bank Contacts Based on Manoeuvring Performance under Environmental Loads

H. Liu, N. Ma & X.C. Gu
Shanghai Jiao Tong University, Shanghai, China

L.P. Wu
Navigation Guarantee Center of North China Sea, Yingkou, China

ABSTRACT: Contacts between container ships and banks or docks in ports pose a great threaten to human life, environment and economy. This paper proposes a method to estimate the probability of ship contacting bank. Big data analysis of ship routes and speeds through the Automatic Identification System and meteorological data considering season wind variation are given to obtain realistic input data for the simulation: the probability distribution of wind, ship routes and ship speeds. At the basis of Maneuvering Modeling Group for manoeuvring motion simulation, the identification of contact is carried out in the channel of Yangshan deepwater port. The Monte Carlo simulation method is applied to obtain a meaningful prediction of the relevant factors of the contact accidents. The proposed approach is demonstrated through the probabilistic analysis for a 10000TEU container ship with different container stacking configurations. The ship speed and stacking strategies in critical meteorological conditions are proposed to avoid the contact events.

1 INTRODUCTION

The Shanghai International Shipping Center Yangshan deepwater port, as the world's largest port to handle container throughput, will reach the throughput of 40 million TEUs per year (People's Daily, July 26th, 2017). The heavy traffic in the port brings in a growing risk of ship-bank contacts, and such risk may end up with much more serious consequent accidents since more and more in-and-out port container ships are holding over 10000 TEUs.

Figure 1 shows the electronic nautical chart of the Yangshan deepwater port. The narrowest section of the waterway to the berths (marked as A) is 1.05km wide, and the width of the approach channel that is close to the dock is no more than 650m. The channel's depth varies from 80m to 16m. Such conditions make a restricted water area for the ships passing through the channel. Meteorological data (Yuan et al. 2002) shows the recorded average and maximum wind speed in the area of Yangshan port are 7.7m/s and 29.1m/s, respectively. So wind force has a non-negligible effect on the ship manoeuvrability. In addition, the ships in close proximity to the bank may suffer from a suction force towards the bank and a yaw moment. This phenomenon is called the ship-bank interaction.

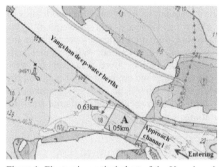

Figure 1. Electronic nautical chart of the Yangshan deepwater port.

Among the research works that deal with the probability of collision, contact and grounding, the number of accident occurrence is currently estimated in two ways: a) by static probability models that are known data in the chain to avoid accidents for Bayesian networks (Hänninen & Kujala, 2012) or fault tree analysis (Obisesan A & Sriramula, 2018), b) by dynamic probability models in which maritime traffic is simulated in the time domain based on data obtained from expert judgement (Uluşçu et al., 2009) or vessel arrival records like Automatic Identification System (AIS) (van Dorp & Merrick, 2011). The first way usually needs empirical probability models to quantify the causation factors (Pedersen, 1995) and defines static critical distance to indicate the occurrence of accidents (Pedersen, 2010). By contrast, the estimation in the second way, e.g. the research by Goerlandt & Kujala (2011) is more realistic because of using AIS data for specific waterways and taking the ship movements in the time domain into account. Furthermore, evasive manoeuvring simulation has been integrated into the probabilistic model (Ståhlberg et al., 2013) in a statistical view rather than a manoeuvring performance based approach. However, although adverse wind condition has been considered as a causation factor (Balmat et al., 2009; Zhang et al., 2013), it has not shown its randomness and dynamic influence on ship motions in the above probabilistic analysis.

The objective of this article is to present a performance-based approach to assess the probability of container ship contact. A 10000TEU container ship passing the narrowest part of the waterway entering Yangshan port, section A as shown in Figure 1, is taken as an example to present the application of the proposed approach. First, the architecture of probability analysis based on the distribution of wind conditions and ship routes through the MCS is proposed. To obtain realistic input data for the simulation, AIS data and record of wind speed and wind direction considering season wind variation are analysed to get the distribution of ship routes and speeds in the channel. Then the manoeuvring equations to identify the candidates of rudder effort insufficiency and consequently the state-space time-domain framework to validate the contact incidents are described. Wind tunnel tests and ship-bank interaction force measurement were carried out to predict the environmental loads on the ship. Finally, the probabilistic analysis of contact are given in terms of ship speeds and container stacking configurations.

2 ARCHITECTURE OF PROBABILITY ANALYSIS

2.1 General outline of probabilistic model

The overall process to predict the probability of contact is outlined in Figure 2. The core methodology is to detect the contact candidates through the MCS. It means that a large numbers of navigation conditions are generated according to a set of probability density functions (PDFs) and the attributes of the ship (main dimensions, speed and load condition). From the replicated simulations of manoeuvring, the incidents of contact are stored. Finally, the program returns sufficient candidates for the risk probability analysis. The criterion for contact is regarded as: whenever the distance between the ship and the bank, y_{bank} is less than the

ship moulded breadth, B (which is the berth-to-dock domain), the contact event occurs.

To increase the efficiency of computing the contact events across the large numbers of MCS runs, a preliminary screening is performed based on the equations of the ship's equilibrium condition. The navigation conditions from which the solution of rudder angle, δ meets the criterion $\delta>35°$ are remained in the candidate group of steering effort insufficiency and further calculated in time domain in the dynamic simulation module. This method will be detailed in section 3.1 and 3.2.

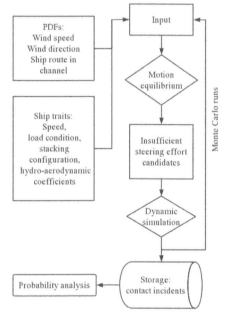

Figure 2. Schematic of probability prediction for contacts.

2.2 Distribution of wind conditions and ship routes

Distributions of wind speed, wind direction and ship-bank distance are investigated prior to the MCS. For the frequency distribution of wind in the area of Yangshan port, the daily wind data recorded by the Shengsi National Level Surface Observational Station from 2012 to 2018 were recounted in terms of speed and direction. These historic files are shared online by China Meteorological Data Service Center (2018). Figure 3 shows the distribution of maximum wind speed and wind direction at maximum speed by different months. Here we choose the distributions in March, July and December to represent the season wind in the spring, summer and winter, respectively. The seasonal variation of predominant wind direction is very clear: the southerly winds in July while the northerly winds in December. The rate of maximum wind speed larger than 15 m/s rises in both December and July. The former is mainly due to frequent cold waves in winter while the latter is because of strong tropical cyclone or even typhoon in summer.

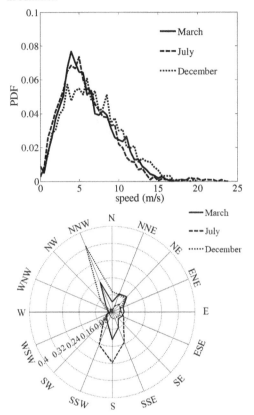

Figure 3. Distributionss of maximum wind speed and wind direction at maximum speed by season in Yangshan port.

The direction of the approach channel is almost the same to the wind direction ESE (for entering ships) or WNW (for leaving ships). The ship is assumed to follow the route parallel to the channel. In this paper, the route distribution of ship entering Yangshan deep-water berths is evaluated from AIS data. The data is shared by the Navigation Guarantee Center of North China Sea (NGCN). The data include 1450 ships from July to December 2018. The distribution of ship route deviation from centerline of the approach channel is shown in Figure 4.

A Fourier series model keeping the first five harmonic terms fits the PDF. The model is defined as Equation 1 and the coefficients of the series model are listed in Table 1.

$$P_{\text{route}} = a_0 + \sum_{n=1}^{5} \left(a_n \cos n\omega x + b_n \sin n\omega x \right) \tag{1}$$

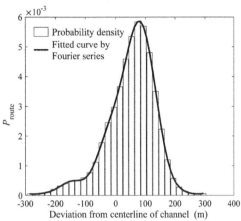

Figure 4. Distribution of ship route deviation from the centerline of the approach channel.

Table 1. Coefficients of the Fourier series model

Terms	Value	Terms	Value
a_0	0.00182	ω	0.01151
a_1	0.00190	b_1	0.00157
a_2	-0.00015	b_2	0.00111
a_3	-0.00034	b_3	8.01E-5
a_4	-4.78E-5	b_4	-0.00019
a_5	1.63E-5	b_5	-6.63E-5

3 MANOEUVRING MODEL FOR CONTACT DETECTION

3.1 Linearization of equation of manoeuvring

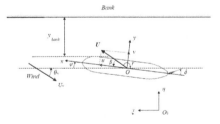

Figure 5. Coordinate systems.

Figure 5 shows the coordinate systems as well as the variables used in the equations of ship motion. The earth-fixed coordinate system O_0-$\xi\eta\zeta$ and ship fixed coordinate system O-xyz are right-handed coordinate systems with the positive ζ and z axis pointing into the page and the origin O at the mid-ship point. The ship initially moves in the direction of the ξ axis with speed, U. U_w and θ_w denote the absolute value of the wind speed and wind direction. Affected by the environmental forces, the ship's velocities are generated as the vector $[u, v, r]$, and the heading angle ψ as well as the drift angle β appears. A rudder deflection δ is required to maintain the ship's direction. y_{bank} is the distance between the ship and the bank.

The model the model of Maneuvering Modeling Group (MMG) for ship manoeuvring by Yasukawa & Yoshimura (2015), To simplify the manoeuvring problem, the ship's engine power adjusts with the wind loads to keep the speed constant, and small deviation in sway and yaw motion is caused by the rudder deflection and wind/bank forces. Therefore, the equation of surge motion is neglected and the non-dimensional equations of ship motion are given as

$$m'\left(\dot{v}' + u'r' + x_G'\dot{r}'\right) = Y_H' + Y_A' \tag{2}$$

$$I_z'\dot{r}' + m'x_G'\left(\dot{v}' + u'r'\right) = N_H' + N_A' \tag{3}$$

where m is the ship mass, I_z is the moment of inertia about the z axis. Y_H and N_H are the hydrodynamic force and moment acting on the ship, and Y_A and N_A are the aerodynamic force and moment due to wind. The non-dimensional treatment are performed based on water density ρ, ship length between perpendiculars L_{PP} and speed U. Some examples are given as follows and readers can get the idea of non-dimensionalization of other variables, too.

$$\dot{v}' = \frac{\dot{v}L_{PP}}{U^2}, \quad \dot{r}' = \frac{\dot{r}L_{PP}^2}{U^2}, \quad u' = \frac{u}{U}, \quad r = \frac{rL_{PP}}{U}$$
$$Y_H' = \frac{Y_H}{0.5\rho L_{PP}^2 U^2}, \quad N_H' = \frac{N_H}{0.5\rho L_{PP}^3 U^2} \tag{4}$$

The hydrodynamic lateral force and yaw moment acting on the ship are expressed by the following polynomial equations

$$Y_H' = Y_v'\dot{v}' + Y_r'\dot{r}' + Y_v'v' + Y_r'r' + Y_\delta'\delta' + Y_\eta'\eta' \tag{5}$$

$$N_H' = N_r'\dot{r}' + N_v'\dot{v}' + N_v'v' + N_r'r' + N_\eta'\eta' + N_\delta'\delta \tag{6}$$

and Y_v', Y_r', N_v', N_r', etc. are the hydrodynamic derivatives on manoeuvring. Y_δ and N_δ are the rudder force derivatives. According to the study of Liu et al. (2016), the ship-bank interaction forces are measurable if $y_{bank} \leq 0.7L_{PP}$. The forces vary with the change of ship-bank distance. To mathematically model this feature, the initial lateral position of the ship is set as $\eta = 0$, and the variation of the forces when η changes is written as $Y_\eta'\eta'$ and $N_\eta'\eta'$. The coefficients Y_η' and N_η' are called asymmetric derivatives.

The aerodynamic lateral force and yaw moment are based on Isherwood (1972)

$$Y_A' = \left(\rho_A/\rho\right)\left(A_Y/L_{PP}^2\right)V_A'^2 C_{YA}\left(\theta_A\right) \tag{7}$$

$$N_A' = \left(\rho_A/\rho\right)\left(A_Y/L_{PP}^2\right)V_A'^2 C_{NA}\left(\theta_A\right) \tag{8}$$

where

$$V_A'^2 = u_A'^2 + v_A'^2 \tag{9}$$

$$u_A' = u' + U_w'\cos\left(\theta_w - \psi\right) \tag{10}$$

$$v_A' = v' + U_w'\sin\left(\theta_w - \psi\right) \tag{11}$$

$$\theta_A = \tan^{-1}\left(v_A'/u_A'\right) \tag{12}$$

and ρ_A is air density, A_Y is the lateral projected area above the waterline, and the wind force coefficients C_{YA}, C_{NA} are wind force coefficients that are expressed as functions of θ_A, the relative wind direction angle.

Yasukawa et al. (2013) considered that when the sway and yaw motion is assumed small, the magnitude of v, ψ and β can be taken as the order $O(\varepsilon)$, and $v' = U'\sin\beta \square \beta$. The following equation can be derived

$$\dot{\eta}' = u'\sin\psi + v'\cos\psi \square \psi + v' \tag{13}$$

Then, Equations (7) and (8) can be rewritten as

$$Y_A' = Y_{A0}' + Y_{A\beta}'\beta + Y_{A\psi}'\psi + O\left(\varepsilon^2\right) \tag{14}$$

$$N_A' = N_{A0}' + N_{A\beta}'\beta + N_{A\psi}'\psi + O\left(\varepsilon^2\right) \tag{15}$$

Readers can refer to (Yasukawa et al., 2012) for the expressions of Y_{A0}', $Y_{A\beta}'$, $Y_{A\psi}'$, N_{A0}', $N_{A\beta}'$ and $N_{A\psi}'$. When the ship reaches the motion equilibrium, the acceleration terms in Eqs. (2) and (3) are eliminated. By taking Eqs. (5)-(8), (14) and (15) into Eqs. (2) and (3), the following equations for the equilibrium condition is obtained

$$0 = Y_{A0}' + \left(Y_v' + Y_{A\beta}' + Y_{A\psi}'\right)\psi_0 + Y_\eta'\eta_0' + Y_\delta'\delta_0 \tag{16}$$

$$0 = N_{A0}' + \left(N_v' + N_{A\beta}' + N_{A\psi}'\right)\psi_0 + N_\eta'\eta_0' + N_\delta'\delta_0 \tag{17}$$

By substituting the wind condition and ship's lateral position into Eqs. (16) and (17), the rudder angle δ_0 and heading angle ψ_0 can be solved.

3.2 Dynamic simulation of contact avoidance

In this paper, once the checked navigation condition falls into the candidates of insufficient steering effort, the dynamic simulation will be carried out. The ship performs conventional stopping and steering action to avoid potential contacts. The stopping time is evaluated based on the regression formula by Nippon Kaiji Kyokai (1979)

$$TIME = 7.17\left(mUC_b/\left(HP\right)^{1/2}\right)^{1/2} \tag{18}$$

where C_b is the block coefficient and HP is the horsepower at U. A proportional-differential (PD) type autopilot is used for course keeping simulations, which is based on the following equation

$$\delta = K_P\left(\psi - T_d r\right) \qquad (19)$$

where K_P and T_d the are a proportional and differential gain constants, respectively. The value of K_P and T_d is tuned by trial-and-error method.

4 CASE STUDY: 10000TEU CONTAINER SHIP

4.1 Target ship

A 10000TEU container ship is studied in this case study. According to the official file of 10000TEU Container Ship Vessel Trim and Stability Booklet, the paper chose a typical non-full loaded condition marked *10T/TEU DEP. AT DESIGN DRAFT* with 7453 containers overall and each of those containers weights 10t. Principle dimensions of the ship and shipping information for the loaded condition are listed in Table 2. The service speed of the ship is V_S=23.75kn.

Following the study by Qiao et al. (2017), five stacking configurations with the same container number are selected to investigate the effect of stacking configuration on the probability of contact. Diagrams and numbering regulation for the five chosen stacking configurations are presented in Figure 6. The paper investigated not only conventional stacking configurations (A1, A2 and A3) also two special on-deck forms like "lack of containers after bridge unit" (A4) and "comb form" (A5).

Table 2. Principle dimensions of the KVLCC2

Specifications	Unit	Full scale
Length btw. perpendiculars L_{PP}	m	320
Moulded breadth B	m	48.2
Design drought T	m	13
Displacement Δ	m^3	124337
Block coefficient C_b	-	0.602
Container number - in hold	-	4578
Container number - on deck	-	2875
Propeller diameter D_P	m	9.5
Rudder area A_R	m^2	72

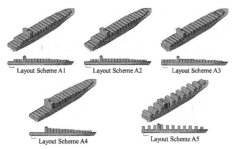

Layout Scheme A1   Layout Scheme A2   Layout Scheme A3

Layout Scheme A4   Layout Scheme A5

Figure 6. Diagrams and numbering regulation for chosen stacking configurations.

Qiao et al. (2017) measured the wind force coefficients, C_{YA} and C_{NA}, through systematic wind tunnel tests in the Wind Tunnel of Shanghai Jiao Tong University. Figure 7 shows the wind load coefficients C_{YA} and C_{NA} for 5 chosen stacking configurations. θ_A = 0, 90, and 180 are termed a "head wind," "beam wind," and "following wind," respectively.

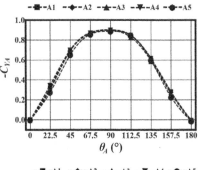

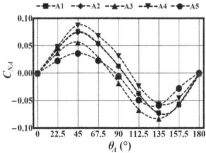

Figure 7. Wind force coefficients for the 10000TEU for five stacking configurations

For hydrodynamic derivatives, systematic planar motion mechanism tests were carried out in the Circulating Water Channel (CWC) of Shanghai Jiao Tong University. The dimensions of measuring section are 8.0m × 3.0m × 1.6m. The PMM tests include oblique towing test, static rudder test, pure sway test and yaw and drift test. Table 4 summarizes the test conditions for all the test types. Moreover, the asymmetric derivatives, Y'_η and N'_η, are analyzed from the asymmetric hydrodynamic test. In this experiment, the ship was placed laterally off the centerline of the CWC with different displacements and water flowed past the ship at speed U (Liu et al. 2017). Table 3 shows the non-dimensional hydrodynamic derivatives.

Table 3. Asymmetric derivatives

Derivatives	Value	Derivatives	Value
$Y'_{\dot{p}_t}$	1.6E-5	$N'_{\dot{p}_t}$	-7.7E-6
$Y'_{\dot{v}}$	-0.0521	$N'_{\dot{v}}$	-0.0001
$Y'_{\dot{p}_r}$	-0.0004	$N'_{\dot{p}_r}$	-0.0002
$Y'_{\dot{y}_t}$	-0.0066	$N'_{\dot{y}_t}$	-0.0038
$Y'_{\dot{y}_r}$	0.0015	$N'_{\dot{y}_r}$	-0.0015
Y'_{δ}	-0.0016	N'_{δ}	0.0007

4.2 Probability of contact

The investigated scenario for the simulation is the process of 10000 TEU passing the narrowest section of the channel, section A as marked in Figure 1. The presented results for each simulation case are obtained based on 10000000 MCS runs. First, the probabilistic simulation is conducted regarding different season wind conditions. The ship speed is 7.13kn corresponding to 30% V_S. Table 4 lists the probabilities of contact for 10000 TEU passing section A in different months. It can be seen that the highest rate of contact appears in July and the rate of contact in March is much lower. The higher risk of contact in July is mainly due to the higher frequency of strong winds in summer. It indicates the consideration of seasonal variation of wind condition in the probability analysis is meaningful.

Table 4. Probability of contacts for 10000TEU passing through section A in different months

Month	Stacking configuration	Probability of contact
March	A1	2.79E-05
	A2	4.76E-05
	A3	2.46E-04
	A4	0.00
	A5	2.47E-04
July	A1	5.83E-03
	A2	5.89E-03
	A3	6.68E-03
	A4	5.50E-03
	A5	6.64E-03
December	A1	1.98E-04
	A2	2.89E-04
	A3	6.25E-04
	A4	5.69E-05
	A5	6.11E-04

Next, the distribution of wind speed and wind direction in July is used and the simulations are conducted with variation in ship speeds and stacking configurations. The rates of contact are shown in Table 6. It can be seen that no contact event is detected in all the stacking configurations at 50% service speed. The maximum probability of contact

appears in the stacking configuration A3, with 0.65‰ at 40%V_S and 6.68‰ at 30%V_S, respectively. The incidence of contact in A5 ranks second, slightly lower than that in A3. The probability of contact is lowest when the ship is stacked in the form of A4.

Table 4. Probability of contacts for 10000TEU passing through section A in July

Ship speed	Stacking configuration	Probability of contact
50%V_S	A1	0
	A2	0
	A3	0
	A4	0
	A5	0
40%V_S	A1	2.03E-04
	A2	2.16E-04
	A3	6.50E-04
	A4	1.59E-04
	A5	5.12E-04
30%V_S	A1	0.00583
	A2	0.00590
	A3	0.00668
	A4	0.00550
	A5	0.00663

4.3 Detailed statistics of contact

A crucial characteristic of the presented simulation model is the ability to provide the detailed information about some of the conditions in which the contact occurs. The wind speed, direction and initial lateral position of the ship can be statistically studied. As an example, Figure 8 to Figure 10 illustrate the probability distribution of the three factors in contacts per stacking configuration at 30% V_S in July. the distribution of wind speed is concentrated in the range of 17m/s-23m/s. In plot (b), the wind directions in contacts falls into the range [-130°, -50°] and [50° 120°]. The two ranges represent a beam wind or a quartering wind to the ship. The distributions of initial lateral positions with respect to the centerline for the five stacking configurations have the trend similar to the distribution of routes for the ships entering Yangshan port as shown in Figure 4, particularly its peak point is very close to the peak of the distribution in Figure 4, with the deviation around $0.25L_{PP}$=80m.

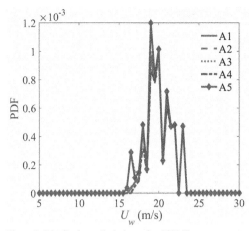

Figure 8. Distributions of wind speed at 30% V_S.

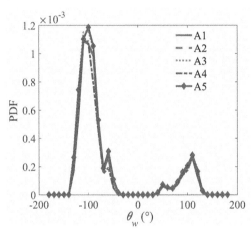

Figure 9. Distributions of wind direction at 30% V_S.

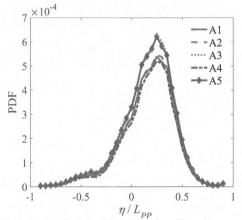

Figure 10. Distributions of initial lateral position with respect to the centerline at 30% V_S.

It can be concluded that the beam wind or quartering wind to the ship with the speed U_w ≥17m/s can be considered as the critical meteorological condition for the container ship when entering the port. The speed should be no less than 50% V_S, so that the probability of contact would be negligible. According to the study by Briggs et al. (2009), the maximum ship speed in a given channel is about 80% of a parameter called the Schijf limiting speed, which is solved based on the formulae in Huval (1980). For the 10000TEU ship, the maximum speed to pass section A is 18.6kn, approximately 80% V_S. So keeping a speed of >=50% V_S is a feasible strategy for the 10000TEU container ship to reduce manoeuvring risks during passing the channel. The stacking configuration also plays an important role in affecting the contact probability of the container ship. A3 and A5 are relatively hazardous stacking configurations in terms of the distributions of wind condition and route in Yangshan port. A4 is a good choice for container ships when avoiding contact in strong wind in the port.

5 CONCLUSIONS

This paper presents a manoeuvring performance based approach for the modelling of contact probability and the application to in the approach channel to Yangshan port is demonstrated. The two major stochastic external factors in causing the risk, wind load and ship-bank interaction are mathematically modelled by PDFs, and the probabilistic trait of them in the contact event are predicted using the MCS technique.

As an example, in the scenario of a 10000TEU container ship entering the narrowest part of the approach channel, simulations are conducted to investigate a) the difference of contact probability due to variation of season winds and b) the influence of ship speed and container stacking configuration on the contact probability. Results show the risk of contact is higher during typhoon season. The advisable ship speed to avoid contact is higher than 50% of the service speed. The type of stacking configuration is also a significant factor in the rate of contact due to strong wind.

The proposed approach is capable of providing detailed information about the circumstances in which the contact takes place. It is expected that the model to evaluate the ship damage extent in a contact accident can be combined, so that the simulation model can be expanded to predict the risk level of ship-bank contact in restricted waterways.

ACKNOWLEDGEMENT

This study is supported by the China Ministry of Education Key Research Project "KSHIP-II Project": No. GKZY010004. Sincere thanks are expressed to Mr Yi Dai, Dr Fei Wang, Mr Dan Qiao and Mr Qiang Chen for their help in the model tests.

REFERENCES

Balmat, J. F., Lafont, F., Maifret, R., and Pessel, N., 2009, Maritime risk assessment (MARISA), a fuzzy approach to define an individual ship risk factor. *Ocean Engineering* 36(15): 1278-1286.

Briggs, M. J., Vantorre, M., Uliczka, K. & Debaillon, P. 2008, Prediction of squat for underkeel clearance. *Handbook of Coastal and Ocean Engineering* 723-774.

China Meteorological Data Service Center, 2017, Dataset of daily surface observation values in individual years in China, http://data.cma.cn/en/?r=data/detail&dataCode=A.0029.0001.

Goerlandt, F. & Kujala, P. 2011. Traffic simulation based ship collision probability modelling. *Reliability Engineering and System Safety* 96(1): 91-107.

Hänninen, M. & Kujala, P. 2012. Influences of variables on ship collision probability in a bayesian belief network model. *Reliability Engineering & System Safety* 102: 27-40.

Huval, C. J. 1980. Lock Approach Canal Surge and Tow Squat at Lock and Dam 17, Arkansas River Project: Mathematical Model Investigation. Technical Report HL-80-17. Vicksburg, MS: U.S. Army Engineer Waterways Experiment Station.

Isherwood, R. M. 1972. Wind Resistance of Merchant Ships, *Transaction of the Royal Institution of Naval Architects* 115: 327-338.

Nippon Kaiji Kyokai. 1979. About ship stopping performance (General Technology). *Transactions of Nippon Kaiji Kyokai* 166: 1-6.

Liu, H., Ma, N., Shao, C. & Gu, X. C. 2016. Numerical Simulation of Planar Motion Mechanism Test and Hydrodynamic Derivatives of a Ship in Laterally Restricted Water. *Journal of Shanghai Jiaotong University* 50(1): 115-122, in Chinese.

Liu, H., Ma, N., and Gu, X. C. 2017. Ship-bank interaction of a VLCC ship model and related course-keeping control. *Ships & Offshore Structures* 12: S305-S316.

Obisesan A, Sriramula S. Efficient response modelling for performance characterisation and risk assessment of ship-iceberg collisions[J]. Applied Ocean Research, 2018, 74:127-141.

Pedersen, P. T. 2010. Review and application of ship collision and grounding analysis procedures. *Marine Structures*. 23(3): 241-262.

People's Daily, July 26th of 2017, http://en.people.cn/n3/2017/0726/c90783-9246682-2.html.

Qiao, D., Ma, N., Gu, X. C., Wang, G. C. & Chen, Q. 2017. Study on energy efficiency design for large container ship with the consideration for the influence of wind load. *Proc. 11th International Conference of Port-city Universities League*: 96-116.

Ståhlberg, K., Goerlandt, F., Ehlers, S. & Kujala, P. 2013, Impact scenario models for probabilistic risk-based design for ship–ship collision. *Marine Structures* 33(5): 238-264.

Tabri, K., Varsta, P. & Matusiak, J. 2010. Numerical and experimental motion simulations of nonsymmetric ship collisions. *Journal of Marine Science and Technology* 15(1): 87-101.

Uluşçu, Ö. S., Özbaş, B., Altıok, T. & İlhan, O. 2009. Vessel traffic in the strait of Istanbul. *Risk Analysis* 29(10): 1454–1472.

van Dorp, J. R. & Merrick, J. R. W. 2011. On a risk management analysis of oil spill risk using maritime transportation system simulation. *Annals of Operations Research* 187(1): 249-277.

Yasukawa, H., Hirono, T., Nakayama, Y. & Koh, K. K. 2012. Course stability and yaw motion of a ship in steady wind. *Journal of Marine Science and Technology* 17(3): 291-304.

Yasukawa, H., Sano, M. & Amii, H. 2013. Wind effect on directional stability of a ship moving in a channel. *Journal of the Japan Society of Naval Architects and Ocean Engineers* 18: 45-53, in Japanese.

Yasukawa, H. & Yoshimura, Y. 2014. Introduction of MMG standard method for ship maneuvering predictions. *Journal of Marine Science and Technology* 20(1): 37-52.

Yuan, Z. X., Min, Q., Tu, H. Y., Zhang, L. G. and Zheng, Q. 2002. Experimental Study on Navigation Safety Simulation of the First Stage Construction of the Yangshan Deep Water Port Area. *Journal of Shanghai Scientific Research Institute of Shipping.* 25(2):64-81, in Chinese.

Zhang, D., Yan, X. P., Yang, Z. L., Wall, A. & Wang, J. 2013. Incorporation of formal safety assessment and bayesian network in navigational risk estimation of the yangtze river. *Reliability Engineering & System Safety* 118(10): 93-105.

Statistical Analysis of the Real Surface Currents and Wind Parameters for the Szczecin Lagoon

L. Kasyk, M. Kijewska & K. Pleskacz
Maritime University of Szczecin, Szczecin, Poland

ABSTRACT: This study focuses on the investigation of the real data on the wind and surface currents parameters for employing them in order to predict the survivor's movement in the Szczecin Lagoon (Polish: Zalew Szczeciński) area. For this purpose, the surface currents parameters were recorded during empirical study on the Szczecin Lagoon. In turn, the wind parameters were measured at the hydro meteorological stations in Świnoujście and Trzebież. The conducted experiments and statistical analyzes of the results allowed to conclude that the movement of surface water masses of the Szczecin Lagoon significantly depends on the incoming air masses over the region. The obtained directions of air masses movement, after their shifting in time by about one and half hours, generally show high compliance with the calculated directions of drifter's trajectories.

1 INTRODUCTION

The leeway has been studied since World War II when Pingree in 1944 reported on the leeway of Navy rafts [Pingree, 1944].

Understanding and creating a model of leeway of the Szczecin Lagoon area may be very important during SAR (search and rescue) actions, or location of pollution contaminants which can be brought in by the Odra River or spilled by ships navigating on the Szczecin Lagoon. The analysis of leeway of the Szczecin Lagoon area based on in-situ experiments has not been carried out so far. In the literature can find works in which some relations between the movement of water masses and the movement of air masses are described.

For instance, in the paper [Hufford and Broida, 1976] , a small craft leeway was determined as a function of the wind speed. The leeway is calculated relative to the surface current by measurement of the separation distance of the small craft from a dyed patch of surface water at sea, using time-sequenced aerial photography. In turn, the surface drifts of two buoys launched in the Adriatic Sea during the DART06 (Dynamics of the Adriatic Sea in Real-Time in 2006) sea trials and in the Liguria Sea during the MREA07 (Maritime Rapid Environmental Assessment in 2007) experiment were predicted in [Vandenbulcke et al., 2009].

However, in the paper [Chang, 2012] the authors received a relation between the observed near-surface current vectors and surface wind vectors for the north-western part of the Pacific Ocean, with strong winds (20-50 m/s).The benefits of high-frequency (HF) radar ocean observation technology for backtracking drifting objects were analyzed. In [Abascal et al., 2012]. In (Cho et al., 2014), an operational search and rescue modelling system was developed to forecast the tracks of victims or debris from marine accidents in the marginal seas of the north-western Pacific Ocean. A stochastic trajectory model was employed within the system to estimate the trajectories of drifting objects. This system is able to consider leeway drift and to deal with uncertainties in the forcing fields. An analysis of a drifter's drift depending on the prevailing wind conditions was carried out for the Gulf of Finland in different seasons of 2011 and 2013 [Delpeche-Ellmann et al., 2016]. The authors proved that the drift of surface waters in the Gulf of Finland can be described by Ekman transport. In [Maio et al., 2016], the most probable search area was estimated with using statistical and deterministic leeway models. The main objective of this paper was to validate a new approach of the leeway calculation that relies on a real person in water event which occurred in the Tyrrhenian Sea in July 2013. In turn, the method for parametrization of surface wind-driven currents in the Black Sea using drifters, altimetry, and wind data was presented in [Stanichny et al., 2016]. The authors showed that their method could be successfully used in order to predict floating objects trajectories basing on the satellite data. In this paper, this method was used for computing the total surface currents and for determining trajectories of oil slick propagation which was observed on satellites images. The drift study of MH370 debris was conducted by means of numerical modelling using a forward particle tracking technique in [Nesterov, 2018]. Four models with respect to the leeway factors and drift angles were considered, including a proposed model of the random distribution of the leeway factors of particles. In turn, in [Dagestad and Röhrs, 2019], satellite derived currents are used for forecasting an ocean trajectory drift. Such trajectories were compared with trajectories based on currents calculated by ocean models. The results of the analysis the correlation of surface water mass movement and air masses movement may be applied to build a model of leeway of a person in the waters of a specific, small, shallow the Szczecinski Lagoon, and consequently, to improve the safety of navigation.

From the point of view of maritime rescue, it is a very important issue, as one of the stages of planning a search and rescue operation is the determination of the position or area of search [IAMSAR, 2013]. Being able to forecast the possible location of a survivor on the basis of the available meteorological data and the developed algorithm of the survivor's leeway should eliminate spontaneous searches and increase the chances of finding a living person. At the end of the 1980s, with the creation of state emergency response areas under the provisions of the International Convention on Maritime Search and Rescue signed in Hamburg on April 27, 1979, it became necessary to develop more effective tools of planning local SAR operations. The Szczecin Lagoon is considered a small and relatively safe reservoir for sailing and motorboats sports. However, the squalls which occur in the area are strong and unexpected as to the direction of the wind impact. Contrary to what may be believed, incidents which require SAR operations on the Szczecin Lagoon are frequent.

The main sources of pollution of the Szczecin Lagoon are the Odra River and spills from ships. The Odra River can transport contaminants from its upper course, from the neighboring agglomerations (Szczecin is the largest city in the Odra Delta estuary, of a population of ca. 370,000) and fields, from the chemical factory Police (a major producer of phosphatise fertilizers) located ca. 20 km north of Szczecin in the area of the estuary of the Odra River [PIOŚ, 1994], and from the conventional coal power station Elektrownia Dolna Odra. Other inflows are relatively insignificant, as they account for about 2% of all tributaries [PIOŚ, 1994]. What is more, the quality of waters of the Szczecin Lagoon is strongly affected by the maritime economy. In 2018, the Szczecin-Świnoujście port complex handled ca. 29 million tons and 75 100 TEU. Vessels of a maximum length of 215 m and draught of 9.15 m can enter the port. Additionally, in the area of the Szczecin Lagoon there are smaller trade ports, such as Nowe Warpno, Police, Przytór, Trzebież or Stepnica, and a number of marinas.

Despite many factors which may in theory affect the movement of waters of the Szczecin Lagoon, short-term variability of its surface currents is essentially related to wind parameters. This property is particularly noticeable on small-sized reservoirs, such as the Szczecin Lagoon [Vandenbulcke et al., 2009]. It is worth noting that the Szczecin Lagoon is a reservoir which does not have any characteristics of an ocean basin. The Szczecin Lagoon is a coastal water reservoir of the flow lake type, intensively flushed with waters of land origin [IMGW, 1980]. Therefore, adopting the circulation of oceanic waters to the conditions occurring in the waters of the Szczecin Lagoon is not the right approach. It is worth mentioning that until now, no comprehensive studies of current parameters have been carried out in the area of the Szczecin Lagoon. Currents are measured only in the Świna Strait, near the Brama Torowa No. 1, and in the vicinity of Trzebież in the Polish part of the Lagoon. However, those linear measurements are taken in a narrow passage (current in or out), depending on the water level, for the purposes of pilotage. What is more, the hydrological conditions prevailing in these regions are different from those in the central (main) part of the Szczecin Lagoon. It is also worth emphasizing that the existing numerical models simplify the movement of surface water masses. They represent the main circulation patterns quite well, yet do not take into consideration local current properties.

Considering the size (surface area and depth) of the Szczecin Lagoon, it is important to take into account local properties of currents. Meanwhile, the existing numerical models, such as, e.g., PM3D or HIROMB, calculate parameters of surface currents for few discrete grid points only (from several to several dozen). Moreover, it's worth adding that there are tools for simulating the transport and behavior of small organisms (microorganisms), e.g., the Legrangian particle-tracking tool. Such tools are based on the assumption that microorganisms, identified in a single sample of water, move passively with currents. In order to be able to verify the results of such simulations, it is worth carrying out empirical research with the use of drifters, and describe patterns of water mass movement (i.e., horizontal diffusion of water masses) on the basis of the results.

The purpose of this study is to statistic analyze the relation between the parameters of leeway and appropriate parameters for the movement of air masses in the selected water reservoir - the Szczecin Lagoon.

Chapter 2 presents the research area, the designed drifters, data obtained and research methods. Chapter 3 presents statistical analysis of drifter's trajectories and wind masses movement parameters in this area. Conclusions are included in Chapter 4.

2 RESEARCH AREA, DATA AND RESEARCH METHODS

The Szczecin Lagoon covers waters of the Odra River estuary (Poland's second largest river) and the southern Baltic Sea. The area is bounded by the following positions: latitude ca. 53°42' N - 53°52' N, longitude ca. 013°53' N - 014°36'. In the north, the islands of Wolin and Uznam separate it from the Baltic Sea. The Lagoon is subdivided into the Large Lagoon (Polish: Wielki Zalew), of a surface area of 488 km², lying within the territory of Poland, and the Small Lagoon (German: Kleines Haff), covering an area of 424 km², which belongs almost entirely to Germany. The southern limit of the Large Lagoon is designated by the Jasienica channel outlet (on the west bank) and the mouth of the Krępa River (in the east).

The Szczecin Lagoon is considered a small and relatively safe reservoir for sailing and motor sports. The major beaches in the Polish part of the Szczecin Lagoon are Stepnica, Trzebież, Czarnocin, Nowowarpieńskie Lake and Wolin. The Lagoon is about 28 km long and over 52 km wide [Locja Bałtyku, 2009]. The danger lies in the shape of the coastline and the bottom, which – in combination with changing hydrodynamic conditions □ have already led to many misfortunes. The bottom in the central part of the Szczecin Lagoon is covered with mud which lies at depths of 4.5 – 5.5 m. The muddy parts of the Lagoon are surrounded by a belt of sandy shallows at a depth of 1 – 1.5 m, which slope steeply towards the muddy bottom [Wolnomiejski and Witek, 2013; Osadczuk, 2004]. The average depth of the Szczecin Lagoon is 3.8 m. The largest natural depth of the Szczecin Lagoon is 8.5 m. Almost 25% of the area of the Szczecin Lagoon has a depth of 0 - 2 m. The greatest depth of 10.5 m is the effect of the deepening of the busy Szczecin-Świnoujście waterway, leading through the Szczecin Lagoon from Szczecin to the Baltic Sea. The Szczecin-Świnoujście waterway, together with the Piastowski Channel, plays an important role in the exchange of waters between the Szczecin Lagoon and the Baltic Sea. The port of Szczecin and the port of Świnoujście form one of the biggest port groups in the Baltic Sea region. What is more, the Piastowski Channel also significantly influences the formation of the unique hydro chemical conditions on the Szczecin Lagoon [Bucholtz, 1986; Majewski, 1958; Majewski, 1964; Majewski, 1972; Maliński, 1968; Robakiewicz, 1993].

The in-situ experiments on the Szczecin Lagoon have been carried in the summer season of 2018 (end of June – mid-October), to describe relations between the parameters of leeway and wind on the Szczecin Lagoon. Ten launches of drifters were performed.

For the purpose of the study, a set of 4 drifters has been designed by scientists at the Maritime University of Szczecin and manufactured by a specialized company. The drifters have balanced buoyancy, so that they do not protrude from the water or are not exposed directly to the wind. At the top of the drifter, directly under the lid, there is a Spot Trace locator, connected to the battery via a power cable. This device measures the geographic position of the drifter with a GPS receiver and sends data in real time via a satellite network. Subsequent positions of the drifter are recorded at 10-minute intervals and then analyzed.

Drifters were deployed in pairs or triples during. Locations were, for each launch, determined individually depending on the wind parameters and with a view to maximizing moving time (due to the costs of the launching operation). Drifters moved over the waters of the Szczecin Lagoon until they got stuck in the reeds by the shore or until they needed to charge the battery.

In this paper, an attempt is made to perform a comparative analysis of trajectories of drifter No 11 (referred to briefly as a drifter later in the article) on the Szczecin Lagoon with wind parameters (direction and speed) measured in the technical basins of the Institute of Meteorology and Water Management (IMGW) in Trzebież - 53° 39'N, 014° 31'E and Świnoujście - 53° 55'N, 014° 15'E. Wind parameters are measured at a height of 10 m. At present, Trzebież and Świnoujście are the closest localities to the Szczecin Lagoon where IMGW performs measurements of wind parameters and makes them available to users. Measurements in the area of Szczecin Lagoon are not performed by IMGW.

First, the basic statistical parameters describing both: during launching and wind conditions prevailing were calculated

The average leeway direction and the average direction of air mass movement were calculated based on [Krawczyk and Słomka, 1982].

$$\overline{\varphi} = \operatorname{arctg} \frac{\sum_{i=1}^{n} w_i \sin \varphi_i}{\sum_{i=1}^{n} w_i \cos \varphi_i} \qquad (1)$$

where: φ_i = the series of directions calculated using readings of the drifter position and directions of the air mass movement; n = the number of measurements registered during one launch; w_i = the weights describing the time between the consecutive measurements of the drifter position.

In this work, due to some unevenness of data registration, additional weights were used, which are sometimes between successive registrations of the drift position.

The average leeway speed and air mass movement speed were calculated using the following formula:

$$\bar{v} = \frac{1}{T} \sum_{i=1}^{n} w_i \cdot v_i \qquad (2)$$

where: v_i = the leeway speed and air mass movement speed obtained between the consecutive measurements of the drifter's position; T = the total drift duration after one launch.

Afterwards, for the description of wind conditions, the following parameters were calculated: dominant wind direction and average wind speed, based on wind directions and speed measured at meteorological stations in Trzebież and Świnoujście. The dominant wind direction is the dominant of measured wind directions during a given launching, while the average wind speed is the arithmetic average of the measured speed.

To determine the compatibility of the trajectories of the drifters' movement with the wind directions, measured in Trzebież and Świnoujście, the following diagrams were made: wind frequency graph, wind rose (the conventional direction of the wind increased by 180 degrees) and the rose of leeway. The compatibility of the trajectories of the drifters' movement with the wind directions in Trzebież and Świnoujście were examined using statistical tests, which checked the significance of this correlation. In this study the Pearson correlation coefficient significance test was used, in which the zero hypothesis is H_0: $\rho = 0$, compared to the alternative hypothesis H_1: $\rho \neq 0$. To verify this hypothesis, the following test statistic was used

$$t = \frac{r}{\sqrt{1-r^2}} \sqrt{n-2} \qquad (3)$$

where: r = Pearson's correlation coefficient; n = the size of the dataset.

The significance of Pearson's correlation coefficient for all drifts was tested. A commonly used scale was used for this purpose: if the Pearson correlation coefficient r < 0.1 - the correlation is very low; if r∈⟨0.1, 0.3) - the correlation is low; if r∈⟨0.3, 0.5) - the correlation is medium; if r∈⟨0.5, 0.7) - the correlation is high; if r∈⟨0.7, 0.9) - the correlation is very high; and r > 0.9 - means nearly perfect correlation. To illustrate the linear relationship between variables, a 95% confidence area for regression line was used.

A similar correlation analysis was performed for leeway and wind speed.

3 COMPARATIVE STATISTIC ANALYSIS OF SURFACE CURRENTS AND WIND PARAMETERS

On the basis the obtained data (geographical coordinates of the drifters' position) from in-situ experiments carried out, the authors calculated directions and speeds of the drifters' trajectories for each time step between one and next registration of position. Based on these data, the average leeway direction and the average speed for each of the ten launches carried out were calculated (See table No 1).

The basic statistical parameters describing the conducted launching are: the average drift direction (The average direction is calculated clockwise from the north in degrees. This direction indicates drifter's trajectory), the dominant wind direction in Trzebież and Świnoujście (direction describing where wind blew in degrees), average drifter's (in m/s) and wind speed in Trzebieży and Świnoujście (in m/s).

Table 1. Basic data on the ten launches carried out in the summer season (end of June - mid-October) 2018

Launch number	Date/time of launching	Drift duration [h]	Average Direction of leeway [°]	The dominant wind direction in Trzebież	The dominant wind direction in Świnoujście	Average leeway speed [m/s]	Average wind speed in Trzebież [m/s]	Average wind speed in Świnoujście [m/s]
1	29.06/16:00	24	221	NNE	NE	0,108	5,53	5,70
2	30.06/16:30	20	207	NNE	NE	0,122	6,16	6,55
3	06.07/16:00	24	114	W	WNW	0,164	4,12	4,18
4	07.07/18:00	39	119	W - NW	W - NE	0,138	3,28	2,93
5	10.07/13:00	47	229	NE*	ENE	0,104	2,45*	5,71
6	13.07/14:00	43	121	NW	NW	0,117	2,59	3,93
7	30.07/19:00	65	239	ESE	NNW	0,106	2,16	3,34
8	16.08/18:00	88	90	SSE - NNE	SSE - W*	0,121	2,37	2,57*
9	06.09/15:00	27	16	SW	SW	0,111	2,14	3,02
10	13.10/13:00	41	322	SSE	SSE	0,154	3,51	3,55

Preliminary analysis of the dominant wind directions in Trzebież and Świnoujście as well as the average directions of drifters' trajectories, included in table 1, allows to state, that in many cases the direction indicating where the wind blows (wind direction increased by 180 degrees) and drifters' trajectories for particular launchings are consistent. In turn, the highest mean leeway speed were obtained for two trials (No 3 - 0.164 m/s and No 10 - 0.154 m/s). It is worth mentioning, that for these two trials, the average wind speeds were not the highest, they were medium and amounted over than 4 m/s for trial No 3 and about 3.5 m/s for trial No 10. The highest average wind speeds were obtained for trials: No 1 - over 5.5 m/s, No 2 - over 6 m/s and for trial No 5 - over 5.5 m/s. The average wind speed can be related to the dominant wind direction.

In the wind from the directions NNE, NE, ENE, average wind speeds were the highest, for other dominating wind directions lower average wind speeds were obtained. Average leeway and wind speeds generally do not correlate with each other. Significantly, for the highest average leeway speeds (0.164 m/s, 0.154 m/s), the highest average wind speeds were not obtained, but medium (4.12 m/s, 4.18 m/s, 3.51 m/s, 3.55 m/s).

For the lowest leeway speed (0.104 m/s, 0.106 m/s, 0.108 m/s), the lowest average wind speed were not obtained, and different averages were obtained (5.71 m/s, 2.16 m/s, 3.34 m/s, 5.53 m/s, 5.70 m/s). For intermediate leeway speed (0.111 m/s, 0.117 m/s, 0.121 m/s, 0.122 m/s, 0.138 m/s), also average wind speed were obtained (2.14 m/s, 3.02 m/s), 2.59 m/s, 3.93 m/s, 2.37 m/s, 6.16 m/s, 6.55 m/s, 3.28 m/s, 2.93 m/s). This means that the average leeway and wind speed calculated for the ten launchings carried out do not correlate with each other directly. The impact on the size of the average leeway speed for a given launch may have other factors, e.g. the area of the Lagoon, where the trial took place.

3.1 Leeway directions vs wind directions

During the research various meteorological conditions prevailed. The wind blew from different directions and had different speeds ranging from close to zero to about 11 m/s. During launchings 1, 2, 3 and 10 the wind from one direction dominated, which translated into a stabilized buoy drift direction (Fig. 1).

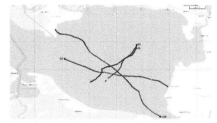

Figure 1. Trajectory No. 1 (d1), 2 (d2), 3 (d3) and 10 (d10), with unchanging direction. The black point is the launching point

For example, for leeway 2, Figure 2 shows the distribution of the frequency of wind directions measured in Trzebież and Świnoujście.

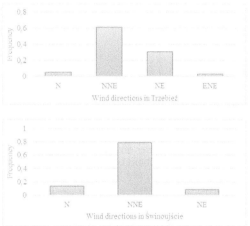

Figure 2. Frequency chart of wind directions for Trzebież and Świnoujście for leeway No 2 (30.06 – 1.07.2018)

Both in Trzebież and Świnoujście, the north-west wind definitely dominated. Below are presented two wind roses (the first - for Trzebież, the second - for Świnoujście) and the leeway rose. It should be borne in mind that the wind roses present the direction in which the wind blows (the conventional direction of the wind increased by 180 degrees). The leeway directions are the directions in which drifter No. 11 was moved.

a)

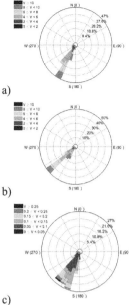

b)

c)

Figure 3. Directional distributions (wind roses) of the wind conditions in the localities: a) Świnoujście, b) Trzebież and c) leeway rose.

Presented wind roses and leeway rose show the compatibility of directions of drift and movement of air masses in Trzebież and Świnoujście. This is also confirmed by the graph of changes in time, the direction of drifter's trajectories and directions of the air masses measured in Trzebież and Świnoujście.

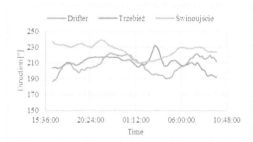

Figure 4. Diagram of changes in trajectories of the drifter No 2 and wind directions measured in Trzebież and Świnoujście.

In turn, in the case of launchings No 8 and 9 (Fig. 5), the meteorological conditions were very variable as to the wind direction.

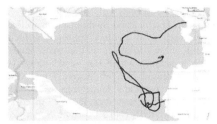

Figure 5. Trajectory No. 8 (d8) and 9 (d9) with variable direction. The black point is the launching point.

For both towns the dominant wind direction is SW. However, the dispersion of existing directions is very high. This is also visible on the following wind roses: for Trzebież and Świnoujście.

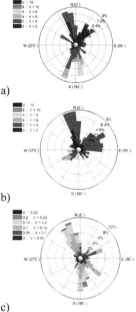

a)

b)

c)

Figure 6. Directional distributions (wind roses) of the wind conditions in the localities: a) Świnoujście, b) Trzebież and c) leeway rose for launching No 9.

Similar scattering of directions and frequencies of their occurrence occur on the drift rose 9, however, the better compliance occurs with the directions of air masses moving in Trzebież. Consistency between leeway and wind direction changes over time is shown in Figure 7.

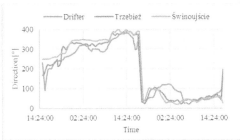

Figure 7. Diagram of changes in trajectories of the drifter No 9 and wind directions measured in Trzebież and Świnoujście.

During the others launchings, two or three phases of stabilization of the leeway direction can be distinguished (No's 4, 5, 6, 7). For example, leeway No 4 was chosen (Figure 8).

Figure 8. Trajectory No 4 (d 4). The black point is the launching point.

For leeway No 4, the graphs of the frequency of wind directions measured in Trzebież and Świnoujście show two groups of dominant directions. For Trzebież, these are the directions SW-W and NW, and for Świnoujście N and W (Fig. 9).

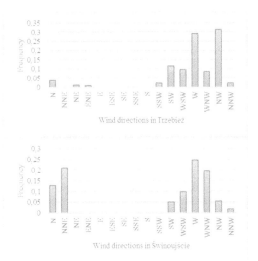

Figure 9. Frequency chart of wind directions for Trzebież and Świnoujście for leeway No 4 (07.07 – 09.07.2018).

The wind roses for Trzebież and Świnoujście show the situation for the directions of the air masses' movement and allow to compare this distribution of directions with the leeway rose No 4. In this case, there is a greater variation in the directions of leeway versus different wind directions.

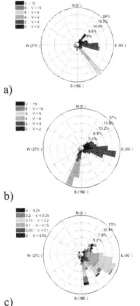

a)

b)

c)

Figure 10. Directional distributions (wind roses) of the wind conditions in the localities: a) Świnoujście, b) Trzebież and c) leeway rose for launching No 4.

On the other hand, consistency over time is presented in the following graph of changes in leeway and wind directions in Trzebież and Świnoujście during trial No 4.

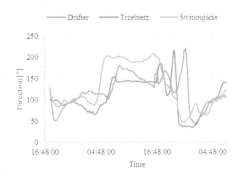

Figure 11. Diagram of leeway and air masses movement directions in Trzebież and Świnoujście, on days 7-07 - 10.07. 2018.

Figures 4, 7, 11 illustrate the general compatibility of leeway directions with the directions of air masses (wind direction increased by 180 degrees) registered in the Trzebież and Świnoujście. To confirm this, statistical tests were performed in the next step, checking the correlation between the direction of the drifters' trajectories and the direction of the air masses.

Chart 11 clearly shows the offset of the leeway direction curve versus the airflow direction curve. Statistically significant correlations were detected comparing drifters' trajectories directions with airflow directions recorded in Trzebież or Świnoujście for some time earlier. The time difference was about 1.5 hours. During the analysis, a strong convergence of the leeway direction with the direction of the air mass movement was found. In each case, in order to determine the significance of the correlation between the direction of leeway and the direction of the air mass movement, the Student's t-test was used and in most cases the test showed the significance of the correlation.

For example, for launching No 2, in Fig. 12 the scatterplot visualizes positive correlation between the leeway direction and the direction of air masses in Świnoujście.

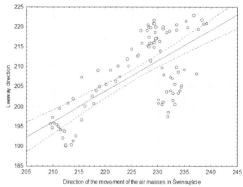

Figure 12. The scatterplot of the directions of air masses movement in Świnoujście and directions of leeway of drifter No 2.

Indeed, the correlation coefficient was 0.684, and the value of the Pearson correlation coefficient value test was 9.15. This value definitely exceeded the critical value, which in this case, for the significance level of $\alpha = 0.05$ and 96 degrees of freedom was 1.66. Thus, the hypothesis about the lack of correlation between the trajectories of drifter No 2 and the direction of the air masses in Świnoujście should be rejected.

In turn, for wind directions recorded in Trzebież during launching drifter No 2, the scatterplot is shown in Fig. 13.

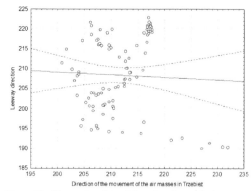

Figure 13. The scatterplot of the directions of air masses movement in Trzebież and directions of leeway of drifter No 2.

In Fig. 13, no correlation can be observed. This is also confirmed by the test statistic of significance of Pearson's correlation coefficient, which amounted to -0.4 and definitely did not fit into the area of rejection of the hypothesis - interval (1.66, ∞).

The significance of the Pearson correlation coefficient for all leeway was tested in an analogous way. Table 2 presents a statistical assessment of the significant correlation between the direction of leeway and the direction of air masses in Trzebież and Świnoujście.

Table 2. Statistical assessment of the significant correlation between leeway direction and wind direction recorded in Trzebież and Świnoujście for particular releases.

Launch number	Compatibility of leeway direction with wind direction in Trzebież	Compatibility of leeway direction with wind direction in Świnoujście
1	high	no significant
2	no significant	high
3	medium	high
4	very high	very high
5	no significant*	very high
6	low	no significant
7	very high	very high
8	nearly perfect	very high*
9	nearly perfect	nearly perfect
10	low	low

* incomplete data

Analyzing table No 2, it can be seen that for a given drifter trajectories, at least for one of the hydro meteorological stations, Trzebież or Świnoujście, the level of correlation is at least high. It is different only for trial No 6 and 10. The common features of these two drifts are: launching area, part is in the breakwater area (Brama Torowa I, Świna) or near estuary the Odra River. These factors could have influenced the results of the correlation of trajectories directions with the directions of air masses movement in Trzebież and Świnoujście. For launching No 6, the low correlation of the directions of this leeway with the directions of air masses in

Trzebież may also result from drifting in the vicinity of the narrows between the Large Lagoon and Small Lagoon. There the drifter made a 360 degree rotation. Meanwhile, the wind directions do not indicate the possibility of rotation.

3.2 *Leeway speed vs wind speed*

Registering a drifter position every 10 minutes allowed for a detailed analysis of calculated leeway speeds. During the whole series of releases, drifters moved at speeds from 0 m/s to 0.348 m/s. The maximum wind speed recorded during the tests was 13.1 m/s in Trzebież and 15.1 m/s in Świnoujście. Among the drift speeds, the most common are values above 0. 1 m/s. For the wind in Trzebież, speeds greater than 4. 5 m/s prevail. And for the wind in Świnoujście average speeds prevail, i. e. from 4 m/s to 7 m/s. The diagram below shows changes in time for leeway speed (in cm/s) and wind speed in Trzebież and Świnoujście (in m/s).

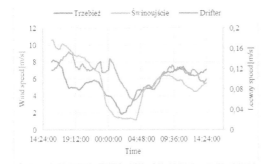

Figure 14. Diagram of speed changes for launching No 1.

As in the case of the direction, also for leeway speed and wind speed, correlations were detected with a time shift of about 1.5 hours (Figure 15).

Figure 15. The wind speed in Trzebież, shifted in time, compared to the speed of leeway for launching No 1.

A significance test of Pearson's correlation coefficient was used to detect the correlation between leeway speed and wind speed.

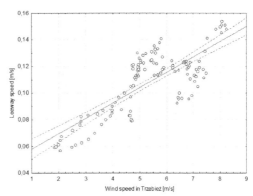

Figure 16. The scatterplot of the speed of the leeway against the wind speed in Trzebież during launching No 1.

The significance value of the Pearson correlation coefficient value is 14.05 and definitely exceeds the critical value, which in this case, for the significance level $\alpha = 0.05$ and 122 degrees of freedom is 1.66. Therefore, the hypothesis about the lack of correlation between the speed of leeway and the speed of winds in Trzebież should definitely be rejected. The correlation coefficient is 0.787. A similar result was obtained by verifying the correlation of the leeway speed with the wind speed in Świnoujście. The correlation coefficient turned out to be significant and amounted to 0.894.

Speeds from all launches were tested in a similar way. The table below presents a statistical assessment of the significant correlation between leeway speed and wind speed recorded in Trzebież and Świnoujście for particular releases.

Table 3. Statistical assessment of the significant correlation between leeway speed and wind speed recorded in Trzebież and Świnoujście for particular releases.

Launch number	Compatibility of leeway speed with wind speed in Trzebież	Compatibility of leeway speed with wind speed in Świnoujście
1	very high	very high
2	high	very high
3	medium	medium
4	high	high
5	not significant*	very high
6	high	high
7	medium	medium
8	high	low*
9	high	medium
10	low	low

* incomplete data

4 DISCUSSION AND CONCLUDING REMARKS

Studies on the movement of surface masses of the Szczecin Lagoon may be of significant importance, e.g. during search and rescue operations or validation of traffic forecasts for these water masses generated by the hydrodynamic model of this area.

In this paper an attempt was made to analyze leeway depending on wind conditions. The main result of statistical research is the discovery, that the movement of air masses significantly influences the movement of surface water masses of the Szczecin Lagoon. Indeed, the directions of leeway in this area for the developed drifter basically depend on the directions of movement of air masses measured in Trzebież or Świnoujście, but shifted in time (delayed) by about one and half hours.

Both under stable wind conditions (wind generally from one fixed direction, during one in-situ experiment), as well as, under variable wind conditions (the change in wind direction or speed), such relationship has been determined. What is more, the significance of the correlation of changes in these directions has also been proven for many experiments and this significance has often been high. In many cases the direction of the leeway was slightly different (from a few to about twenty degrees) from the direction of air masses in Trzebież or Świnoujście, and sometimes it diverged to the left and sometimes to the right from the direction of air masses movement. At wind speeds of at least 2-3 m/s, the direction of the air masses movement was more stable, and thus the direction of the leeway was also stable. At wind speeds of no more than 2-3 m/s, the directions of the leeway and the movement of air masses were changing, but these changes were statistically significant (which was confirmed by the analysis of the significance of their correlation).

In addition, changes in the speed of the leeway and wind also proved statistically significant in many cases. A change in speed of the wind is often followed by change in leeway speed. But the delay in change of the leeway speed in relation to the speed of air masses is about 1.5 hours.

This research outcome has been achieved under research project No 1/S/INM/18 financed with a subsidy from the Ministry of Science and Higher Education for statutory activities.

REFERENCES

Abascal, A. J., Castanedo, S., Fernández, V., Medina, R., 2012. Backtracking drifting objects using surface currents from high-frequency (HF) radar technology. *Ocean Dynamics* 62. 1073-1089.

BHMW (Biuro Hydrograficzne Marynarki Wojennej). Locja Bałtyku. Wybrzeże Polskie (502), 2009. Wydanie IX.

Bucholtz, W., 1986. Charakterystyka hydrologiczno-hydrodynamiczna rejonu Zalewu Szczecińskiego. IM Oddział Szczecin.

Chang, Y.C., Chen, G.Y., Tseng, R.S., Centurioni, L.R., Chu, P.C., 2012. Observed near-surface currents under high wind speeds. *Journal of Geophysical Research* 117. Art. No. C11026.

Cho, K.-H., Li, Y., Wang, H., Park, K.-S., Choi, J.-Y., Shin, K.-I., Kwon, J.-I., 2014. Development and Validation of an Operational Search and Rescue Modeling System for the Yellow Sea and the East and South China Seas. *Journal of Atmospheric and Oceanic Technology.* 31. pp. 197-215.

Dagestad, K.-F., Röhrs, J., 2019. Prediction of ocean surface trajectories using satellite derived vs. modelled ocean currents. *Remote Sensing of Environment* 223. pp. 130-142.

Delpeche-Ellmann, N., Torsvik, T., Soomere, T., 2016. A comparison of the motions of surface drifters with offshore wind properties in the Gulf of Finland, the Baltic Sea. Estuarine, *Coastal and Shelf Science* 172. Pp. 154-164.

Hufford, G.L., Broida, S., 1976. Estimation of the leeway drift of small craft. Ocean Engineering 3(3). pp. 123-132.

IAMSAR MANUAL, International Aeronautical and Maritime Search and Rescue Manual, Mission Coordination, II.IMO/I CAO. London.

IMGW (Instytut Meteorologii i Gospodarki Wodnej), 1980. Zalew Szczeciński. Praca zespołowa pod redakcją Aleksandra Majewskiego. Wydawnictwa Komunikacji i Łączności. Warszawa.

Krawczyk, A., Słomka, T., Podstawowe metody modelowania w geologii. Materiały pomocnicze do ćwiczeń. AGH. Kraków. 1982.

Maio, A., Martin, M.V., Sorgente, R., 2016. Evaluation of the search and rescue LEEWAY model in the Tyrrhenian Sea: a new point of view. *Natural Hazards and Earth System Sciences* 16. pp. 1979-1997.

Majewski, A. ,1958. Hydrologiczne badania Zalewu Szczecińskiego 1952-1957. *Gazeta Obserwatora PIHM* 7.

Majewski, A., 1964. Ruchy wód Zalewu Szczecińskiego. *Prace PIHM* 69.

Majewski, A., 1972. Charakterystyka hydrologiczna estuariowych wód u polskiego wybrzeża. *Prace PIHM* 105.

Maliński, J., 1968. Wyznaczenie ruchu wody w Świnie, Pianie i Dziwnie oraz zmian położenia powierzchni Zalewu Szczecińskiego. Oddział Morski IMGW, Gdynia.

Nesterov, O., 2018. Consideration of various aspects in a drift study of MH370 debris. *Ocean Science* 14. pp. 387-402.

Osadczuk, A., 2004. Zalew Szczeciński: środowiskowe warunki współczesnej sedymentacji lagunowej. Wydawnictwo Naukowe Uniwersytetu Szczecińskiego. *Rozprawy i studia* T. 549

Pingree F. W., 1944. Forethoughts on Rubber Rafts. Woods Hole Oceanographic Institution, 26 pp.

PIOŚ (Państwowa Inspekcja Ochrony Środowiska), 1994. Zalew Szczeciński. Wielki Zalew. Zmiany jakościowe w wieloleciu. Praca zespołowa pod redakcją Tadeusza Mutko. *Biblioteka Monitoringu Środowiska*, Warszawa.

Robakiewicz, W., 1993. Warunki hydrodynamiczne Zalewu Szczecińskiego i cieśnin łączących Zalew z Zatoką Pomorską. Monografia pod redakcją W. Robakiewicza. Gdańsk.

Stanichny, S.V., Kubryakov, A.A., Soloviev, D.M., 2016 Parameterization of surface wind-driven currents in the Black Sea using drifters, wind, and altimetry data. *Ocean Dynamics* 66: p. 1–10

Vandenbulcke, L., Beckers, J.-M., Lenartz, F., Barth, A., Poulain, P.-M., Aidonidis, M., Meyrat, J., Ardhuin, F., Tonani, M., Fratianni, C., Torrisi, L., Pallela, D., Chiggiato, J., Tudor, M., Book, J.W., Martin, P., Peggion, G., Rixen, M., 2009. Super-ensemble techniques:Application to surface drift prediction. *Progress in Oceanography* 82. pp. 149-167.

Wolnomiejski, N., Witek, Z., 2013. The Szczecin Lagoon Ecosystem: The Biotic Community of the Great Lagoon and its Food Web Model. Versita Ltd., London, 293 pp.

Inland, Costal, Port and Offshore Innovations

A Novel Algorithm for Modeling Human Decision Making of Inbound Merchant Ships - A Case Study of the Shanghai Waigaoqiao Phase IV Port

J. Xue
Delft University of Technology, Delft, The Netherlands
Wuhan University of Technology, Wuhan, China
National Engineering Research Center for Water Transport Safety (WTSC), Wuhan, China

C.Z. Wu
Wuhan University of Technology, Wuhan, China
National Engineering Research Center for Water Transport Safety (WTSC), Wuhan, China

P. van Gelder
Delft University of Technology, Delft, The Netherlands

ABSTRACT: With the continuous development of large-scale, high-speed and professional ships, and the increasing construction of modern intelligent deep-water ports, the safety of inbound merchant ships receives more and more attention. Thus, it is of great significance for the stakeholders to address the safety management and realize the intelligent decision-making of the inbound merchant ships properly. In this paper, a novel algorithm for modeling human decision-making of inbound merchant ships is proposed. This method can be used to realize the automatic acquisition and representation of the crew's decision-making knowledge in inbound merchant ships analysis. To verify the performance of the model, a case study based on this method is conducted in the Waigaoqiao Phase IV Port of Shanghai. The experimental results indicate that the maneuvering decision recognition model combined with the method of classification interval division, which proposed in this article, possesses a high reasoning speed and can accurately and scientifically standardize the boundary of the interval of influencing factor data and identify current maneuvering behavior. The proposed methods and the evaluation results provide useful insights for effective safety management of the inbound merchant ships.

1 INTRODUCTION

Currently, waterway transportation plays an increasingly important role in cargo transportation. It accounts for 95% of total crude oil transportation and 99% of total iron ore transportation. However, with the increasing number of vessels and the increasingly busy routes, the environmental pollution related to waterway transportation, the high labor costs and the lack of safety have also received more attention (Lun et al., 2016). In addition, with the development of technologies, such as computers science, communications, artificial intelligence, internet of things, and information physics systems, have greatly advanced the process of ship intelligence and made unmanned autonomous ships a possibility. Autonomous ship technology has developed rapidly in recent years. However, there are still many problems need to be solved. In addition, the existing research does not form a set of theoretical methods to solve the problem of autonomous learning of the autonomous

merchant ship for the maneuvering decision-making characteristics of chew.

At the same time, water transportation is recognized as a high-risk industry. According to statistics, in ship collision accidents, 89% to 96% of accidents are caused directly or indirectly by human factors, and one of the important ways to solve ship accidents caused by human factors is to utilize autonomous maneuvering of ships. In addition, the number of crews is declining recently, while the wages of crew are rising year by year, which has become the second largest expenditure item after the fuel costs of shipping (Lun et al., 2016). As autonomous ships have outstanding advantages in improving the safety management, energy consumption management, and operational efficiency of ships, therefore, the researches for autonomous ships have become an inevitable trend for future ship development and gained the interest of many researchers in both academia and private sectors (Goerlandt & Montewka, 2015).

Classification is a data mining (DM) technique used to predict or forecast the unknown information using the historical data. Many classification algorithms have been developed such as Decision tree (Cohen, Rokach, & Maimon, 2007), Classification and regression tree (Friedman, Olshen, & Stone, 1984), Bayesian classification (Heckerman, 1998), Neural networks (Rojas, 1996) and K-nearest neighbor classification (Denœux, 1995), etc. Among them, the decision tree has become more popular algorithm as it has several advantages over others.

Common decision tree algorithms are Iterative Dichotomiser 3 (ID3), C4.5, C5.0, Classification And Regression Trees (CART), CHi-squared Automatic Interaction Detector (CHAID), etc. In these algorithms, the ID3 algorithm is the influential and wide used decision tree generation algorithm (Jin, Li, & Li, 2014; Umanol et al., 1994; Xiaohu, Lele, & Nianfeng, 2012). It chooses the attribute with the highest information gain as the test attribute of the current node. It divides the sample set based on the value of the test attribute, how many different values of the test attribute exist, the number of subset divisions, and then further divides the corresponding subset of the sample using a recursive method. The C4.5 algorithm is complex when continuously processing data, and its workload is large. C5.0 mainly adds support for Boosting, which uses less memory and is more accurate, but C5.0 is a commercial software, and the public cannot easily get the source code (Witten, Frank, Hall, & Pal, 2016). CART uses the training set and the cross-validation set to continuously evaluate the performance of the decision tree to prune the decision tree, thus achieving a good balance between training error and test error. However, CART and CHAID only supports building binary trees, while ID3 and C4.5 allows two or more outcomes and supports binary or multifork trees (Wu et al., 2007). Therefore, this paper utilizes the ID3 algorithm to learn the crew's maneuvering decision characteristics considering the advantages of it, thus to construct a human-like decision-making model under multiple constraints in a specific scenario.

In summary, this study focuses on the concept of human-like maneuvering for the autonomous merchant ships and studies the human-like decision-making method of for the autonomous chemical tanker. By establishing autonomous learning method of maneuvering decision-making, the maneuvering decision-making rules of typical maneuvering style is explored, and the processes of autonomous learning crew's maneuvering decision-making characteristics for autonomous ships are studied, and the autonomous ship human-like decision-making model is constructed. This study provides a new perspective and methodology for the development of autonomous ship technology in theory and in practice and promotes the application and spreading of autonomous merchant ships.

2 METHODOLOGY

2.1 *Decision tree classification algorithm*

The machine learning technique for inducing a decision tree from data is called decision tree learning, or decision trees. In decision theory and decision analysis, a decision tree is a graph or model of decisions and their possible consequences, including chance event outcomes, resource costs, and utility. It is a method to solve complex decision problems through tree-like logical thinking and can be used to create a plan to reach a goal. Decision trees are constructed in order to help with making decisions. A decision tree is a special form of tree structure and a descriptive means for calculating conditional probabilities.

Decision tree learning is a common method used in data mining. Each internal node corresponds to a variable. A leaf node represents a possible value of target variable given the values of the variables represented by the path from the root node. A tree can be "learned" by splitting the source set into subsets based on an attribute value test. This process is repeated on each derived subset in a recursive manner. The recursion is completed when splitting is either non-feasible, or a singular classification can be applied to each element of the derived subset. Figure 1 shows the flowchart for tree-based classification.

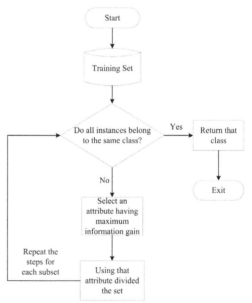

Figure 1. General tree-based classification

Decision tree-based classification is also a most widely used classification method in the field of data mining. The goal is to create a model that predicts the value of a target variable based on several input variables. Each internal node corresponds to one of the input variables. Each leaf node represents a value of the target variable given the values of the input variables represented by the path from the root node to the leaf node. Some of the key advantages of using decision trees are the ease of use and overall efficiency. A decision tree can be represented as a set of production rules in the form of IF-THEN. Each root-to-leaf path in the decision tree corresponds to a rule. And the rule set can be derived that are easy to interpret.

2.2 The proposed model

The ID3 algorithm is a typical decision tree learning algorithm. It uses the information gain method as the attribute selection criterion to determine the appropriate attributes to be used when generating each node. In this way, the attribute with the maximum information gain can be selected as the test attribute of the current node, so that the information required for classification using the training sample subset obtained by the attribute is minimized.

Shannon proposed the information theory in 1948 (Shannon, 1948), and the amount of information on events could be calculated as follows:

$$I(S_i) = -p(S_i)\log_2 p(S_i) \tag{1}$$

where $p(S_i)$ is the probability of occurrence of event S_i.

Suppose that there are v mutually exclusive events $S_1, S_2, ..., S_v$, and only one of them happens. The average amount of information can be measured as follows:

$$I(S_1, S_2, ..., S_v) = -\sum_{i=1}^{v} p(S_i)\log_2 p(S_i) \tag{2}$$

when $p(S_i)=0$, then $I(S_i) = -p(S_i)\log_2 p(S_i) = 0$.

Assume that D is the intelligent ship human-like decision-making training data set contains a set of m classes, $|D|$ stands for the total number of samples in data set D, and $|S_i|$ is the number of samples in data set D that belongs to class $S_i (i = 1, 2, ..., m)$. If we randomly select a sample from D, and this sample belongs to class S_i, then we can get a prior probability of the event as follows:

$$p_i = |S_i|/|D| \tag{3}$$

The expected information (also referred to as entropy) needed to classify D into m classes is defined as:

$$I(|S_1|, |S_2|, ..., |S_m|) = -\sum_{i=1}^{m} p_i \log_2(p_i) \tag{4}$$

Suppose a feature/attribute A has n distinct values, $\{a_1, a_2, ..., a_n\}$, feature/attribute A partitions D into n subsets, $\{D_1, D_2, ..., D_n\}$, $|D_j|$ is the number of samples in subset $D_j (j = 1, 2, ..., n)$, and $|S_j^i|$ stands for the number of samples in subset D_j that belongs to class S_i. Then, the expected information is defined as:

$$E(A) = \sum_{j=1}^{n} \frac{|D_j|}{|D|} I(|S_j^1|, |S_j^2|, ..., |S_j^i|) \tag{5}$$

Note that the smaller the entropy value is, the higher the purity of the subset partition, where m for a given subset D_j,

$$I(|S_j^1|, |S_j^2|, ..., |S_j^i|) = -\sum_{i=1}^{m} p_{ij} \log_2(p_{ij}) \tag{6}$$

The information gain of feature/attribute A is expressed as follows:

$$Gain(A) = I(|S_1|, |S_2|, ..., |S_m|) - E(A) \tag{7}$$

A good rule of thumb would seem to be to choose that attribute to branch on which gains the most information. ID3 examines all candidate attributes and chooses A to maximize $Gain(A)$, then forms the tree and then uses the same process recursively to build decision trees for the residual subsets.

3 EXPERIMENT

We collect the operational data from the unlimited navigational class seafarers based on the simulator of Navi-Trainer Professional 5000, which conforms to the IMO STCW78/10 convention and the Det Norske Veritas (DNV) from the Maneuvering Simulator Laboratory Waterway Road Traffic Safety Control and Equipment Ministry of Education Engineering Research Center. The scenario is the Shanghai Waigaoqiao wharf Phase IV Port, and the ship was downstream of the berthing into the port. We use a merchant ship of 30,000-ton bulk carrier as our experimental ship OS1 (33089.0 t, 182.9 meters long, 22.6 meters wide).

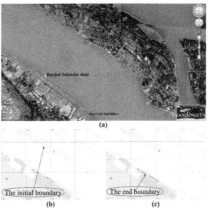

Figure 2. The designed experimental scenario

We use the experienced crew (unlimited navigational class, skilled maneuvering level, captain/chief officer). The multisource information of ship maneuvering traffic environment were collected. For instance, the location (longitude, latitude), environment (wind, current, etc.), control (rudder order, marine telegraph), ship movement (heading, roll rate, etc.), the ship's draft, tugs, mechanical contact force-related parameters, and other related parameters. The above factors, such as the environment, the control, location and the relevant parameters of the tug and other factors were selected from the weakly related parameters.

In order to let the maneuvering decision-making knowledge to be automatically obtained and expressed along with higher decision-making knowledge effectiveness. It is typically necessary to divide the number of linguistic terms by experience (Yuan & Shaw, 1995). In this paper, experimental data of each maneuvering decision-making factor are trisected into three levels, namely, small (a), medium (b), and large (c). To objectively describe the characteristics of each influencing factor, and make it easier to describe how each factor influences final maneuvering decisions. We select six environmental influence factors as the input of our proposed model to study the decision-making mechanisms for different maneuvering behaviors: Current direction (Y1), current speed (Y2), relative current direction (Y3), relative wave direction (Y4), relative wind direction (Y5), relative wind speed (Y6). Table 1 lists some of the training samples.

The crew maneuvers the ship by operating different telegraph and rudder orders to change ship's speed and direction and to complete the ship's control, the symbol are shown in Table 2. The standardization principle of the input selected influence factors for inbound maneuvering decision-making can be seen from Xue et al. (2019). Moreover, Table 3 shows the combining telegraph and rudder orders and the standardization principle for output maneuvering decision-making factors. For instance, X1 (U1D1U2T2) indicates that the maneuvering decision-making is: {Keep the propeller forward and keep the current rudder angle-port rudder}.

Table 1. Training samples for evaluation of the studied area (partially)

No.	X		Y1	...	Y6
	Rudders Order	Telegraphs Order			
1	-1.070	-30.000	318.400	...	7.724
2	-1.767	-30.000	318.400	...	7.724
3	-2.000	-30.000	318.400	...	7.723
4	-2.000	-30.000	318.400	...	6.704
5	-2.000	-30.000	318.400	...	6.704
6	-2.000	-30.000	318.400	...	6.704
7	-2.000	-30.000	318.400	...	6.704
8	-2.000	-30.000	318.400	...	6.685
9	-2.000	-30.000	318.400	...	6.685
10	-2.000	-30.000	318.400	...	6.685
11	-2.000	-30.000	318.400	...	6.685
12	-2.000	-30.000	318.400	...	6.685
13	-2.000	-30.000	318.400	...	6.685
...	...		...	...	...

Table 2. The symbol of telegraph orders (speed control) and rudder orders (course control)

Attributes	Speed control Symbolic principle	Status	Symbol
Variety	$a_{i+1} - a_i \neq 0$	Changed	C1
	$a_{i+1} - a_i = 0$	Unchanged	U1
Direction	$a_i \geq 0$	Ahead	D1
	$a_i = 0$	Stop engine	M1
	$a_i < 0$	Astern	T1

Attributes	Course control Symbolic principle	Status	Symbol
Variety	$b_{i+1} - b_i \neq 0$	Changed	C2
	$b_{i+1} - b_i = 0$	Unchanged	U2
Direction	$b_i \geq 0$	Starboard	D2
	$b_i = 0$	Midships	M2
	$b_i < 0$	Port	T2

Table 3. Maneuvering decision-making factors and standardization principle (output)

Maneuvering Decisions factors	Symbols	Decisions	symbol
X	U1D1U2T2	X1	C1M1C2T2	X19
	U1T1U2T2	X2	C1M1C2D2	X20
	U1D1U2D2	X3	U1M1C2T2	X21
	U1T1U2D2	X4	U1M1C2D2	X22
	C1D1C2T2	X5	C1M1U2T2	X23
	C1T1C2T2	X6	C1M1U2D2	X24
	C1D1C2D2	X7	U1D1U2M2	X25
	C1T1C2D2	X8	U1T1U2M2	X26
	U1D1C2T2	X9	C1D1C2M2	X27
	U1T1C2T2	X10	C1T1C2M2	X28
	U1D1C2D2	X11	U1D1C2M2	X29
	U1T1C2D2	X12	U1T1C2M2	X30
	C1D1U2T2	X13	C1D1U2M2	X31
	C1T1U2T2	X14	C1T1U2M2	X32
	C1D1U2D2	X15	U1M1U2M2	X33
	C1T1U2D2	X16	U1M1C2M2	X34
	U1M1U2T2	X17	C1M1U2M2	X35
	U1M1U2D2	X18	C1M1C2M2	X36

4 RESULTS AND DISCUSSION

4.1 *Standardizing of training set*

The data in Table 1 are standardized according to the principle of standardization of maneuvering decision influence factors proposed by Xue et al. (2019) and Table 3; then we get the training set as shown in table 4.

Table 4. Training set with the principle of standardization (partially)

No.	X	Y1			...	Y6		
		a	b	c	...	a	b	c
1	X10	0	0	1	...	0	1	0
2	X10	0	0	1	...	0	1	0
3	X10	0	0	1	...	0	1	0
4	X2	0	0	1	...	1	0	0
5	X2	0	0	1	...	1	0	0
6	X2	0	0	1	...	1	0	0
7	X2	0	0	1	...	1	0	0
8	X2	0	0	1	...	1	0	0
9	X2	0	0	1	...	1	0	0
10	X2	0	0	1	...	1	0	0
11	X2	0	0	1	...	1	0	0
12	X2	0	0	1	...	1	0	0
13	X2	0	0	1	...	1	0	0
...	...				...	....		

4.2 *Constructing the decision tree*

In the ID3 decision tree algorithm analysis, approximately 80% of the data is randomly selected as the training set, and the remaining 20% is used as the test set. Then, through the proposed model in section 2, we could obtain the decision tree structure, as shown in Figure 3.

Then we can get the decision-making rule set based on the decision tree structure in the form of IF-THEN. Each path from the root node to the leaf node constitutes a rule. For instance, we can get the rule from the left side of the tree structure: *IF Y2=a AND Y3=a AND Y6=a THEN X=X1.* The characteristics of the internal nodes of the path correspond to the conditions of the rule, and the classification of the leaf nodes corresponds to the conclusion of the rule. As a result, we can easily extract the human-like decision-making knowledge using the decision tree and rule set.

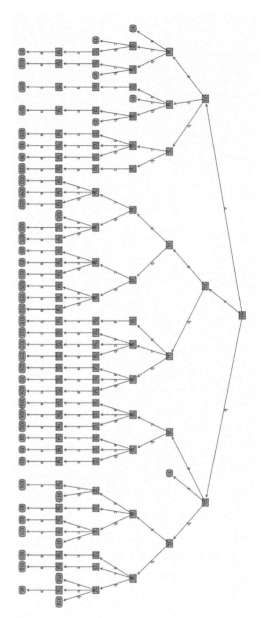

Figure 3. The decision tree structure

5 CONCLUSIONS

With the development of modern science and technology, the improvement of intelligent ships has been technically feasible. The purpose of this research is to recognize the automatic acquisition and representation of the crew's decision-making knowledge and to provide a basis and reference for the development of decision-making algorithms for intelligent ships. In this study, the standardization principle of ship maneuvering is introduced, and we proposed a novel model for the learning of human-like decision-making mechanisms of intelligent ships. This research provides a new perspective for the development of intelligent ship and promotes the application and promotion of intelligent ships in specific scenario. In the follow-up study, we will pay attention to the detailed standardization principles of various influencing factors and maneuvering decision-making factors, and the application of our proposed model for different scenarios.

ACKNOWLEDGMENTS

This study is supported by the National Nature Science Foundation of China (51775396), the Major Project of Technological Innovation of Hubei Province (2016AAA007, 2017CFA008), and the China Scholarship Council.

REFERENCES

Cohen, S., Rokach, L., & Maimon, O. (2007). Decision-tree instance-space decomposition with grouped gain-ratio. *Information Sciences, 177*(17), 3592-3612.

Denœux, T. (1995). A k-Nearest Neighbor Classification Rule Based on Dempster-Shafer Theory. *Systems Man & Cybernetics IEEE Transactions on, 25*(5), 804-813.

Friedman, J. H., Olshen, R. A., & Stone, C. J. (1984). Classification and regression trees. *Belmont, CA: Wadsworth & Brooks.*

Goerlandt, F., & Montewka, J. (2015). Maritime transportation risk analysis: Review and analysis in light of some foundational issues. *Reliability Engineering & System Safety, 138*, 115-134.

Heckerman, D. (1998). A tutorial on learning with Bayesian networks. In *Learning in graphical models* (pp. 301-354): Springer.

Jin, C., Li, F., & Li, Y. (2014). A generalized fuzzy ID3 algorithm using generalized information entropy. *Knowledge-Based Systems, 64*, 13-21.

Lun, Y. V., Lai, K.-h., Wong, C. W., & Cheng, T. (2016). Introduction to Green Shipping Practices. In *Green Shipping Management* (pp. 3-15): Springer.

Rojas, R. (1996). Neural Networks-A Systematic Introduction Springer-Verlag. *New York.*

Shannon, C. E. (1948). A mathematical theory of communication. *Bell system technical journal, 27*(4), 379-423.

Umanol, M., Okamoto, H., Hatono, I., Tamura, H., Kawachi, F., Umedzu, S., & Kinoshita, J. (1994). *Fuzzy decision trees by fuzzy ID3 algorithm and its application to diagnosis systems.*

Witten, I. H., Frank, E., Hall, M. A., & Pal, C. J. (2016). *Data Mining: Practical machine learning tools and techniques*: Morgan Kaufmann.

Wu, X., Kumar, V., Ross Quinlan, J., Ghosh, J., Yang, Q., Motoda, H., . . . Steinberg, D. (2007). Top 10 algorithms in data mining. *Knowledge and Information Systems, 14*(1), 1-37.

Xiaohu, W., Lele, W., & Nianfeng, L. (2012). An Application of Decision Tree Based on ID3. *Physics Procedia, 25*, 1017-1021.

Xue, J., Wu, C., Chen, Z., Van Gelder, P., & Yan, X. (2019). Modeling human-like decision-making for inbound smart ships based on fuzzy decision trees. *Expert Systems with Applications, 115*, 172-188.

Yuan, Y., & Shaw, M. J. (1995). Induction of fuzzy decision trees. *Fuzzy sets and systems, 69*(2), 125-139.

Polar Plot Capability of a Tug in Indirect and Direct Mode of Escort Towing

J. Artyszuk
Maritime University of Szczecin, Szczecin, Poland

ABSTRACT: The mechanism of getting a steady-state (equilibrium) condition of a tug performing escort towing in the whole range of speed is discussed. This is based on a simple but powerful analytical model. Both indirect and direct mode are investigated. Untypically, polar plots of the towing force envelope are finally computed separately for each of them, revealing some interesting specific features. Thus the user (tug designer or master) can have at hand their individual contributions to the overall performance. It is found that both modes are surprisingly important in the escort and complement each other very well. Moreover, the indirect mode, at least in steady-state, may lose its effectiveness at moderate to low speeds in that it goes with pull much below the bollard pull. A rapid transition to the direct mode can solve this operating problem.

1 INTRODUCTION

The port tugs are designed to render assistance, referred to as control or towing, to ships in port areas, like approach channels and basins. The assisted ships, mostly larger ones, are here of limited manoeuvrability in relation to available space, environmental conditions, and requirements for their controlled movement. This towing activity, in terms of the assignment and the later actual (vs. just pure standing-by) usage of tugs, mostly takes place for normal/routine operations of a ship, but shall account for any emergency (e.g. her propulsion or steering failure) that can occur. The latter aspect contributes to the operational risk management.

For certain economic and safety reasons the assistance (by pulling/pushing) must be brought up at high speed, in particular in approach channels. In such situations, while a ship is sailing under her own power, tugs (or a tug) are often put in stand-by, even on a line, and acting if necessary. They work at a ship's s side or stern, thus applying a backing and/or steering force. Such an operation is often called 'escort', which implies a few other related terms sometimes used, e.g. 'escort mode', 'escort tug'. The latter, however, may also denote this purpose-built tug and class thereof (DnV GL, 2018). They work in escort mode, mostly connected (but not only) with the so-called indirect towing. In the latter there is taken an advantage, more or less prominently, of the hydrodynamic lateral force developed on the underwater hull of a tug. The hull force may contribute to a much higher pull than the bollard pull

of the tug. According to various sources, the thruster works little in the indirect towing, more for providing just the turning moment to keep the angular position of a tug. The angular dynamics or statics, in contrast to linear (force) balance, of the tug appears to be hardly a challenge.

The tug performance in escort operation can be partly assessed/compared through the so-called polar (or butterfly) diagram of the maximum effective pull force, which a tug can develop. Its value is plotted therein for each towing direction and escort speed, and naturally presented on a compass rose. Despite a great popularity of escort towing and a huge volume of related research, there is however a significant scatter and/or lack of polar capability data. This unfavourable situation in publishing touches upon many aspects, for example: the associated tug's hydrodynamic data (hull drift and thruster angles, hull and thruster forces), methods of assessment and analysis for polar plots, systematic investigations and sensitivity analysis of various tug design options.

For some practical application, the polar plots are generally presumed to reflect the steady-state (equilibrium) conditions of a tug. They can be directly determined through model tests, full-scale tests (DnV GL, 2018, 2016), computation or real-time man-in-the-loop simulation.

The latter approach is adopted in (Hensen, 2003, pp. 59-61), which incomprehensibly criticises the computational method of seeking for an equilibrium, whereas the obtained 'practical' solution might be not optimal (subject to error). The computational approach is well encouraged by e.g. (Brandner & Renilson, 1994) with a provision of some hydrodynamic input data. These references provide meaningful, readable and rather reliable polar plots in both direct and indirect mode of escort towing. They can serve some partial benchmarking or validation of other results. However, the quoted here charts are apparently quite different from the results of (Hutchison & Gray, 1993).

The Author's previous study (Artyszuk, 2014) dealt in general with some basic relationships involved in the indirect towing. The present report provides further results for this mode in that it collects, processes and generalizes all data required to establish the polar plot. Throughout this research, a contribution to the theory of tug equilibrium is thus expected. Its understanding can enhance a better designing and operating a tug engaged in escort operation. For completeness and consistency of the knowledge, the present study also covers and compares the other mode, of direct towing.

2 METHODOLOGY

As aforementioned, the detailed model presented in (Artyszuk, 2014), including simple hydrodynamic input data, is hereafter used and constitutes an inextricable connection with this paper. Some four sequential fundamental relationships are defined therein, see eqs. (14), (13), (15), (17) (marked as 1° to 4°) of that reference. The symbols adopted are recalled in Figure 1. The first equation returns $\cot(\delta) = f_1(\beta, \gamma)$ – cotangent of thruster angle, lying in the range (-180°, +180°), as function of drift and hawser angle. The thruster angle δ, by means of the unique inverse function (arc tangent), has to be then determined in the drift angle domain. Based on that, the other two equations consistently provide F_T and F_{yH} (the towing force and the lateral hull hydrodynamic force), both expressed as ratio of thruster force F_P, thus referred to as relative forces. In turn, F_P is given by the last fourth equation, when the escort speed is settled.

However, some little computational challenge, rather of mathematical/numerical than physical nature, is required with the first fundamental equation. Namely, since the arc tangent is not unique and basically returns values of δ from (−90°, +90°), that should be properly transposed to the whole 360° range: (−180°, +180°). Otherwise, the negative thruster force F_P has to be allowed, which would be against the initial assumptions. However, the basic physical criterion for the equilibrium is that the resulting absolute/dimensional towing force be positive (which for a negative thruster force, in view of the above, implies a negative relative towing force). This means pulling the hawser (vs. false

'pushing') or having the hull lateral force direction corresponding to the drift angle.

A discretisation every 15° in hawser angle is generally adopted in the computations and presentations, which is however reduced down to 5° in some instances of research. Although dual thrusters (propellers) are installed by standard on a tug, they are supposed in the model to operate in parallel for simplifying the equilibrium solution. This way the highest resulting thruster force is enabled, since both thrusters can act as (or be replaced with) a 'single unit' of doubled power.

For generating the dimensional data of tug performance, of equal importance to dimensionless (universal) ones, specially pertaining to the actual escort speed, an exemplary yet typical tug is assumed. Her particulars are: length abt. 30 m, underwater lateral area of abt. 153 m², bollard pull 50 t (metric tons). The latter is meant to be maximum thrust of propellers with no detrimental effect of advance speed as commonly known from the propeller hydrodynamic characteristics.

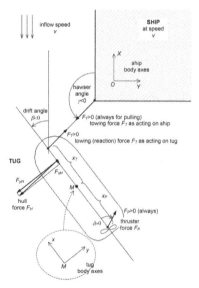

Figure 1. Escort towing geometry and symbols.

The quoted former study (Artyszuk, 2014) was significantly dealing with the indirect towing as being exactly in the equilibrium (or steady-state) condition. The feature of it is that a tug, see Figure 1, sets her starboard side against the inflow ($\beta < 0$). If a tug is getting parallel with the assisted ship ($\beta = 0$) and starting next to expose her port side ($\beta > 0$), whatever the hawser angle, the direct towing is involved. However, the naming of the latter mode is more conventional than physical, since in some cases the hull hydrodynamic force has also some contribution to the effective hawser pull. The math model of Artyszuk (2014) is also fully ready to cover this interesting case. In the example of Figure 1 (a tug on port quarter of assisted ship), the indirect mode is thus encompassed by the drift $\beta \in (-90°, 0°)$, while the direct mode covers $\beta \in (0°, +90°)$.

3 FURTHER RESULTS ON INDIRECT TOWING

Within this chapter, some properties of (Artyszuk, 2014) findings will of course be implicitly reminded to some extent too, but often set in different, more generalised form and focus in view of the stated objectives. However, many new essential relationships on the equilibrium mechanism of indirect towing have emerged.

Figure 2 presents the relative towing force as the ratio of thruster force, $F'_T = F_T/F_P$, in the drift angle domain, being fundamental to the unique solutions of the equilibrium. The hawser angle is the parameter to this family of curves. The whole plot is here independent of the escort speed. As one can notice, there exist some local maxima and multiple drift angles (tug's angular position vs. the flow) for the towing force of equal 'effectiveness'. Although not shown in the present paper, the pattern of hull lateral force of the tug expressed in terms of the thruster force, $F'_{yH} = F_{yH}/F_P$, is surprisingly very similar to that in Figure 2. Nevertheless, Figure 3 illustrates, for more clarity, the inherent features of the towing force related to the hull force, F_T/F_{yH} ($= F'_T/F'_{yH}$). Based on (Artyszuk, 2014, Fig. 3), the thruster angle is a (unique) function of the drift angle for each hawser angle.

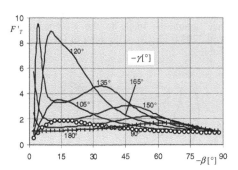

Figure 2. Relative towing force (as ratio of thruster force) in drift domain for various hawser angle.

If we choose in Figure 1 the local maximum for each hawser angle, and next the maximum value ('global maximum') equal either to the above or the value on the (left) interval limit, whichever is higher, then the maximum relative towing force can be plotted against the hawser angle. This is portrayed in Figure 4, as x-y plot, or in Figure 5, as polar (butterfly) plot.

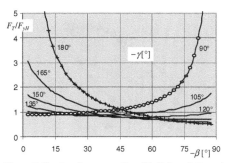

Figure 3. Towing force as ratio of hull lateral force in drift domain for various hawser angle.

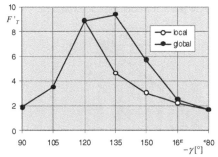

Figure 4. Maximum relative towing force vs. hawser angle.

Of course, Figure 5 demonstrates only one quarter of the overall capability of the tug, or one half of her stern deployment (as per Figure 1), in that the other (right-hand) one is symmetrical. Figures 3 and 5 can potentially serve a tug designer in optimisation process.

If we want to go towards the absolute values (in tons) of towing force (Artyszuk, 2014), first we need to establish the escort speeds of interest, being often in the maximum order of 8-10 knots, i.e. abt. 4-5 m/s. Then we get the following charts in Figure 6, based on hawser angle −120° (providing a single solution), or in Figure 7 associated with hawser angle −175° (giving almost the backing force). The latter is also very interesting, almost extreme case, where even 3 solutions are likely – we are looking therein for an equilibrium, which secures the highest towing force. The also computed cases for 1 and 2 m/s are not presented in Figure 6, since they are similar to the case 3m/s, yet with proper scale, as one can visually extrapolate from the trend shown. If F_P is above 50 t (the assumed reference maximum value for thruster force), then the technically available maximum absolute towing force must be reduced. This reading has been demonstrated in the middle part of Figure 6, on the example of speed 4 m/s. However, especially at lower speeds, if the thruster force needed for equilibrium is much below 50 t, the case 3 m/s, then the towing force would be often less than the nominal 50 t.

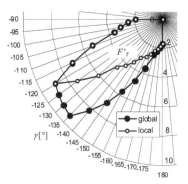

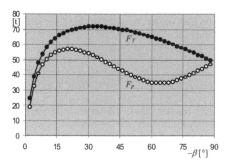

Figure 5. Polar plot of maximum relative towing force.

Figure 7. Absolute towing and thruster forces at 5 m/s – a representative example 2 (hawser angle $\gamma = -175°$).

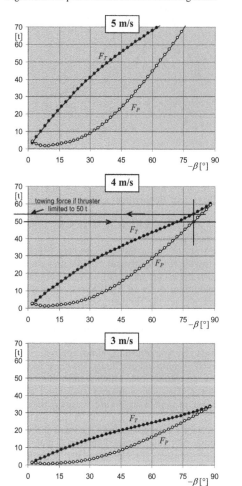

Applying the above mentioned procedure for various speeds and hawser angles, the overall polar capability of a tug in indirect mode of escort towing is arrived at – Figure 8 (very recognisable chart in tug industry yet of limited information and usage). Details are given in Table 1 and 2, in particular comprising data on necessary drift β and thruster angle δ to keep the equilibrium.

It is noticeable in Table 1 that very high towing forces, followed by thruster forces (as well as hull lateral forces) of the same order, are being found at some instances of higher escort speeds and hawser angles. As mentioned, these allowable maximum values for the equilibrium can not be reached due to the actual thruster limit. The values to be adjusted are underlined and marked bold in Table 1. The thruster angle missing in Table 2 can be partly taken from the previous Table 1, valid only at original towing force values, i.e. not being subject to the adjustment. For the other, adjusted values a reference should be made to the aforementioned (Artyszuk, 2014, Fig. 3), with input of the new drift from Table 2.

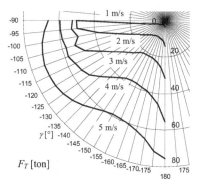

Figure 6. Absolute towing and thruster forces at selected escort speeds – a representative example 1 (hawser angle $\gamma = -120°$).

Figure 8. Polar plot of maximum absolute towing force for various escort speeds.

Table 1. Directly derived absolute towing forces (without thruster force limit)

| | | | escort speed | | | | | | | |
| | | | 5 m/s | | 4 m/s | | 3 m/s | | 2 m/s | |
$-\gamma$ [°]	$-\beta$ [°]	δ [°]	F_T [t]	F_P [t]	F_T [t]	F_P [t]	F_T [t]	F_P [t]	F_T [t]	F_P [t]
180	2	180	**87**	**82**	**56**	**52**	31	29	14	13
175	32	171	**72**	**53**	46	34	26	19	11	8
170	42	160	66	38	42	25	24	14	11	6
165	48	145	62	30	40	19	22	11	10	5
160	54	122	60	25	38	16	22	9	10	4
155	60	95	59	25	38	16	21	9	9	4
150	66	74	59	31	38	20	21	11	9	5
145	74	60	60	42	39	27	22	15	10	7
140	88	50	**63**	**61**	41	39	23	22	10	10
135	88	45	**68**	**66**	44	43	25	24	11	11
130	88	40	**75**	**73**	48	47	27	26	12	12
125	88	35	**83**	**81**	**53**	**52**	30	29	13	13
120	88	31	**94**	**93**	**60**	**59**	34	33	15	15
115	88	26	**110**	**109**	**70**	**70**	40	39	18	17
110	88	21	**133**	**132**	**85**	**85**	48	48	21	21
105	88	16	**171**	**170**	**109**	**109**	**61**	**61**	27	27
100	88	11	**240**	**240**	**154**	**153**	**86**	**86**	38	38
95	88	7	**409**	**409**	**262**	**262**	**147**	**147**	**66**	**66**
90	88	2	**1430**	**1430**	**915**	**915**	**515**	**515**	**229**	**229**

Table 2. Adjusted absolute towing forces (thruster force limited to 50 t)

| | escort speed | | | | | | | | | |
| | 5 m/s | | 4 m/s | | 3 m/s | | 2 m/s | | 1 m/s | |
$-\gamma$ [°]	F_T [t]	$-\beta$ [°]	F_T [t]	$-\beta$ [°]	F_T [t]	$-\beta$ [°]	F_T [t]	$-\beta$ [°]	F_T [t]	$-\beta$ [°]
180	**76**	**46**	**55**	**14**	31	2	14	2	3	2
175	**72**	**36**	46	32	26	32	11	32	3	32
170	66	42	42	42	24	42	11	42	3	42
165	62	48	40	48	22	48	10	48	2	48
160	60	54	38	54	22	54	10	54	2	54
155	59	60	38	60	21	60	9	60	2	60
150	59	66	38	66	21	66	9	66	2	66
145	60	74	39	74	22	74	10	74	2	74
140	**63**	**78**	41	88	23	88	10	88	3	88
135	**65**	**74**	44	88	25	88	11	88	3	88
130	**68**	**72**	48	88	27	88	12	88	3	88
125	**70**	**66**	**52**	**86**	30	88	13	88	3	88
120	**71**	**64**	**54**	**80**	34	88	15	88	4	88
115	**72**	**60**	**56**	**74**	40	88	18	88	4	88
110	**72**	**56**	**58**	**80**	48	88	21	88	5	88
105	**72**	**52**	**58**	**64**	**53**	**84**	27	88	7	88
100	**70**	**48**	**59**	**60**	**50**	**76**	38	88	10	88
95	**68**	**44**	**59**	**56**	**53**	**72**	**50**	**85**	16	88
90	**66**	**40**	**59**	**52**	**52**	**66**	**50**	**80**	**50**	**87**

Based on Figure 8, it is clear that indirect towing (in view of the adopted definition) is effective at hawser angles corresponding to nearly perpendicular tug's deployment. For lower escort speeds, e.g. 1-3 m/s, the curves of this polar plot are rapidly 'getting flat' in the horizontal direction, i.e. when a tug moves towards the stern of the assisted ship. The expected shapes are here to be more or less circular (butterfly-like). According to these curves, the indirect mode paradoxically produces the towing forces much lower than the reference 50 t. Even at the stern location ($\gamma = 180°$), the towing forces are really very low. This is in contrast to the apparent experience on a tug. However, the tug must then change over to the direct mode, see the next chapter.

A special future concern shall be devoted to remarkable differences (and potential inherent relationships) between the polar plots of relative (Figure 5) and absolute (Figure 8) towing forces.

NOTE: In the present study, if someone wishes to quickly change over his/her mind to escort speed in knots, it is enough to multiply twice the values in m/s

4 NEW RESULTS ON DIRECT TOWING

By analogy to the previous case of indirect towing (in terms of methodology, analysis content, and data visualisation), the following equivalent charts for the direct towing have been developed. The direct mode numerically features positive drift angles. In the arrangement of Figure 1, this means a tug having exposed her port side to the inflow, unlike the starboard side actually shown in that figure (as originally associated with indirect mode).

Consequently, Figures 9 and 10 present the relative towing and hull force as the pure output of the equilibrium algorithm. Both are different than in the indirect mode. Much range of Figure 10 is negative, which make the equilibrium physically impossible. That is why the derived Figure 11 has been narrowed/clipped to the positive range as well, being of practical importance.

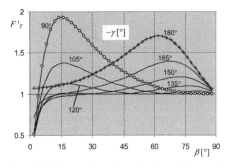

Figure 9. Relative towing force (as ratio of thruster force) in drift domain for various hawser angle - DIRECT TOWING.

In Figure 11 there is no curve for hawser angle −90° (no equilibrium). This is because of the drift angle range considered in these preliminary computations (max. +90°). Normally for this perpendicular position of a tug, the drift must increase little more than +90°, so the tug would be moving then a little her stern first. Of course, one can also try the previous indirect mode, as shown in Table 2.

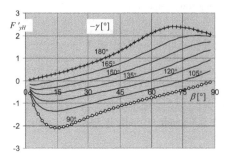

Figure 10. Relative hull lateral force (as ratio of thruster force) in drift domain for various hawser angle - DIRECT TOWING.

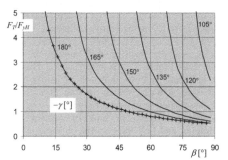

Figure 11. Towing force as ratio of hull lateral force in drift domain for various hawser angle - DIRECT TOWING.

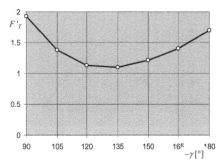

Figure 12. Maximum relative towing force vs. hawser angle - DIRECT TOWING.

The maximum relative towing force in the direct mode (although based only on local maxima) is also above 1, but less than 2, which was the minimum for the indirect mode – compare Figures 12 and 13 to the former Figures 4 and 5.

In the direct mode a different relationship is exhibited between the thruster angle and the drift angle, Figure 14. The reader may refer to the mentioned (Artyszuk, 2014, Fig. 3) for the indirect mode data.

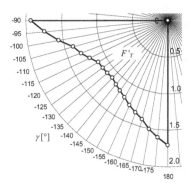

Figure 13. Polar plot of maximum relative towing force - DIRECT TOWING.

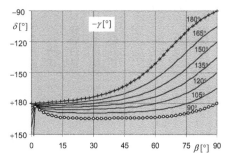

Figure 14. Thruster angle vs. drift angle for various hawser angle - DIRECT TOWING.

The mechanism of looking for the maximum feasible absolute towing forces is also different in some aspects from that of indirect towing – Figure 15. First of all, there appears a region of drift angle providing no equilibrium, starting from zero to a certain vertical asymptote. This was also roughly mentioned before, while commenting on Figure 10. The position of this discontinuity (of physical nature), producing for the equilibrium extremely high towing and thruster forces (with the order of a few hundreds to a few thousands in many cases, especially at high speeds), is associated with the hawser angle and equals $(180° + \gamma)$. Namely, in the example of Figure 15, it corresponds to $\beta = 45°$. If $\gamma = -100°$, for instance, the equilibrium range would amount the last 10° to the right side of the considered drift domain (from 80° to 90°). The asymptote means a tug nearly in line with a hawser, with just a small deflection (high sensitivity) towards inflow. But this is not the only equilibrium position, one may have another one – see the middle/lower case of Figure 15. In general, the towing force fairly accurately 'follows' the thruster force. Under the limit to the latter, the maximum towing force equals just that limit (50 t in our case).

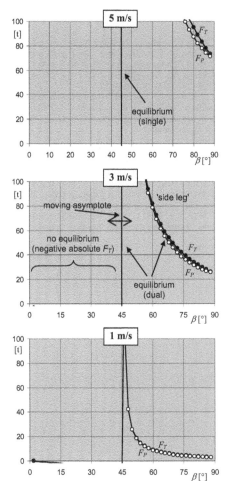

Figure 15. Absolute towing and thruster forces at selected escort speeds – a representative example (hawser angle $\gamma = -135°$) - DIRECT TOWING.

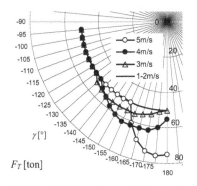

Figure 16. Polar plot of maximum absolute towing force for various escort speeds - DIRECT TOWING.

However, there are some exceptions to this convergence, especially for hawser angle close to −180° and high escort speeds, where the equilibrium should be taken from the 'side leg'. The latter might ensure the towing force much higher than the thruster force.

The whole steady-state situation with the direct towing is presented in Fig. 16.

One can even attempt to superimpose both Figures 8 and 16, which would now present the overall performance of a tug. This qualitatively agrees with the cited foreign results.

5 CONCLUSIONS

The present study has revealed a certain complexity of escort towing performance of an omni-directional tug in steady-state condition, in indirect and direct mode as well. The focus has been made on the physical and computational mechanism leading to the polar diagram capability (Figures 8 and 16) at various escort speeds. The results seem beneficial for both a tug designer and a tug master. In the latter context, many hydrodynamics- or control-related details, in terms of the drift and thrust angles, and the resulting pull force are not available onboard a tug. The tug control becomes then intuitive, requiring much experience and skills, and often being less effective, especially if transient manoeuvring effects are next included.

A noticeable fact is that in both modes an augmentation of effective pull vs. bollard pull is observed, although in different regions, with different intensity, and with different flexibility in tug's control. The most striking finding is that the indirect mode must be immediately given up if the

the applied towing force in steady-state is much below the bollard pull, or a tug becomes unstable. The moment of necessary 'direct⇔indirect' transition may often be unknown. Both modes might also mix each other and be confused.

In addition, against initial expectations, it has been proved that the high thruster force is very essential in keeping the equilibrium in the indirect mode.

A simple yet quite general and adequate model has been adopted for the study. How the results are sensitive to this model and to the assumed tug particulars is a challenging future issue, especially on a systematic basis. Firstly, the hull hydrodynamics may be refined to some actual data from scale experiments or CFD calculations. Secondly, some neglected effects – e.g. restraints on heel angle and engine load, the well-known influence of advance speed on thruster performance – may be implemented. However, all these extra elements will impose additional (negative) restraints on the polar capability thus reducing it.

One of potential deficiencies of the model could also be the parallel operation of dual propulsors. Their independent operation (with the essential effect of their lateral spacing) can excite an arbitrary combination of the total force (although thus reduced) and the yaw moment. The free control of the latter can be looked at as introducing to our mechanical system an 'additional' couple of forces. In the adopted approach the magnitude of moment goes proportional to $F_P \cdot \sin\delta$.

REFERENCES

Artyszuk, J. 2014. Steady-state manoeuvring of a generic ASD tug in escort pull and bow-rope aided push operation. *TransNav, the International Journal on Marine Navigation and Safety of Sea Transportation*, vol. 8, no. 3 (Sep): 449-457. DOI: 10.12716/1001.08.03.17.

Brandner P. & Renilson M. 1994. Hydrodynamic Aspects of Shiphandling Tugs. In: *3rd International Conference - Manoeuvring and Control of Marine Craft (MCMC'94), Sep 7-9*. Southampton.

DnV GL 2016. *Full scale testing of escort vessels (class guideline)*. DNVGL-CG-0155, ed. Feb. Hovik: DnV GL.

DnV GL 2018. *Rules for classification: Ships*. DNVGL-RU-SHIP/Pt.5 Ch.10 Sec. 11 (*Tugs and escort vessels*, pp. 99-117), ed. July. Hovik: DnV GL.

Hensen H. 2003. *Tug Use In Port, A Practical Guide*. 2nd Ed., London: The Nautical Institute.

Hutchison B.L. & Gray D.L. 1993. New Insights into Voith Schneider Tractor Tug Capability. *Marine Technology*, vol. 30, no. 4 (Oct): 233-242.

Use of Association Rules for Cause-effects Relationships Analysis of Collision Accidents in the Yangtze River

B. Wu, J.H. Zhang & X.P. Yan
Wuhan University of Technology, Wuhan, China

T.L. Yip
Hong Kong Polytechnic University, Hung Hom, Kowloon, Hong Kong

ABSTRACT: In order to discover cause–effect relationships in collision accidents, an association rule-based method is applied to analyze the historical accident data in the Jiangsu section of the Yangtze River from 2012 to 2016. First, the Apriori algorithm is introduced for interesting rules mining, and three types of measures of significance and interest are considered, which are support, confidence and lift. Second, the data are discretized based on previous studies and work experience, and the R software is introduced to facilitate the modeling process. Third, the contributing factors are discovered from the cause-effect relationship analysis. Finally, the generated rules are visualized using the Gephi software to further analysis the unknown relationships and patterns. The observed patterns of collision accidents can be avoided by cutting off some factors in the sequential chain of collision accidents, which is beneficial for prevention of collision accidents. Consequently, this paper provides a data-driven method for accident analysis and prevention.

1 INTRODUCTION

Ship collision is a typical accident and accounts for a lot among all maritime accidents (Cai et al. 2017, Sedova et al. 2018). For example, from 1953 to 2002, 461 serious maritime accidents occurred in the Istanbul Straits, the majority of them, i.e. 45.34%, were ship collisions (Akten 1999). In the Gulf of Finland, collision accidents rank second from 1997 to 2006 (Kujala et al. 2009). From 2008 to 2013, the collision accidents accounted for 65% Tianjin Port (Zhang et al. 2018). Similarly, in the Jiangsu section of the Yangtze River, the ship collision accidents account for more than 60% (Jiang 2010).

Owing to the relatively high occurrence of ship collision accidents, many studies focused on ship collision risk analysis and mitigation. Wang & Yang (2018) proposed a novel method to calculate the severity of water transport accidents based on Bayesian network. Sii et al. (2001) developed a fuzzy logic-based model by considering various variables in the concept design stage. Sedova et al. (2018) presented a fuzzy neural system for collision avoidance in busy waterways. Moreover, Hu et al. (2007) proposed a safety assessment method for risk management of waterborne transport after defining four unique criteria.

From previous studies, valuable insights have been gained. However, it can be seen that the previous models require the practitioners have good knowledge and understanding of the accident development, which may cause some uncertainty. However, association rule is a data mining method, which means it manages to discover the patterns of collision accidents only from the accident data without any prior knowledge. Therefore, the motivation of this paper is to use association rules to discover the cause-effect relationships from a variety of causation factors.

The remainder of this paper is organized as follows. The definitions of association rules are introduced in Section 2. The cause-effect relationship analysis of ships collision accidents is introduced in Section 3, which includes data sources, association rules of ship collision accidents, analysis of high support and lift, and visualization of association rules. Conclusions are drawn in Section 4.

2 ASSOCIATION RULES

2.1 *Introduction of association rules*

Association rule learning is a rule-based machine learning method for discovering hidden relationships between variables in a database from the perspective of data mining. When introducing it to ship collision accident analysis, after discovering the patterns of ship collision, it is meaningful to take countermeasures to cut off the necessary causation factors in an association rule. For example, an association rule for ship collisions in the Yangtze River is {accident area = anchorage} => {encounter scenarios = collision with stationary ship}. It shows that the collision with stationary ships will have a large probability to occur when ships are anchored at an anchorage. Therefore, the officer on watch should always take sharp lookout to prevent the occurrence of collision accidents when anchoring in the anchorage.

2.2 *Definition of association rules*

In association rules, the collision data record set is called the database, defined as *D*; the collection of all items is called the itemsets, which is defined as *I*; each accident record in the database is defined as *T*, where $T \in D$. The set of items that appear simultaneously in accident record is called an itemset, and it is defined as *K-th* itemset. The items on the left and right of the symbol "=>" are referred to the antecedent and the succedent of the association rules. The frequency that uses to measure the occurrence frequency of an item is defined as *support*.

$$support\left(A \Rightarrow B\right) = P\left(A \cap B\right) \qquad (1)$$

when *A* belongs to the data record *D*, the probability of *B* also belongs to *D* is defined as *confidence*, i.e. the conditional probability.

$$confidence\left(A \Rightarrow B\right) = P\left(B|A\right) = P\left(A \cap B\right) / P\left(A\right) \quad (2)$$

Another method used for measuring the relationship is called *lift*, which is defined as the interest (Ochin et al. 2016, Grabot 2018).

$$lift\left(A \Rightarrow B\right) = confidence\left(A \Rightarrow B\right) / P\left(B\right)$$
$$= \left(P\left(A \cap B\right)\right) / \left(P\left(A\right)P\left(B\right)\right) \qquad (3)$$

From this definition, if lift = 1, *A* and *B* are independent and have no influence; if lift > 1, *A* and *B* are interdependent, mutually reinforcing and positively correlated; if lift < 1, *A* and *B* are mutually constrained, mutually reinforcing and negatively correlated (Barati et al. 2017). Moreover, if lift = 0, *A* and *B* will not occur simultaneously.

2.3 *Association rule mining*

There are three widely used algorithms for association rules, which are Apriori algorithm, Partition-based algorithm and FP-tree algorithm, respectively. Apriori algorithm, proposed by Rakesh & Ramakrishnan in 1994 (Srikant & Agrawal 1994), is a basic algorithm in association rules and widely used as a classical algorithm (Borah & Nath 2018, Weng & Li 2017, Xu et al. 2018). This paper uses this algorithm because it is intuitive, concise and easy to implement. The flow chart of Apriori algorithm is shown in Figure 1.

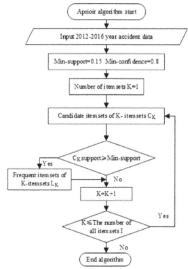

Figure 1. Flow chart of Apriori algorithm

3 CAUSE-EFFECTS RELATIONSHIPS OF INLAND SHIPS COLLISION ACCIDENTS

3.1 *Data Sources of ship collision accidents*

The ship collision accidents data were collected by Jiangsu Maritime Bureau. The database fully records the collision information of ships accidents in inland waterway from 2012 to 2016. The original collision data are discretized, and then the following set of accident features are obtained. The dataset contains the following information: the Yangtze River water period (i.e. flooding season, normal season, and dry season), time (i.e. daytime and nighttime), location of accidents (i.e. Changshu, Changzhou, Jiangyin, Nanjing, Nantong, Taicang, Taizhou, Yangzhou, Zhangjiagang and Zhenjiang), accident area (i.e. channel, anchorage and others), severity of consequences (i.e. negligible, minor and major), shipwreck (i.e. yes or no), number of fatalities (i.e. 0 fatalities, 1-2 fatalities and 3-9 fatalities),Causation factors (i.e. ship conditions, environmental factor, human factors and cargo factors), encounter scenarios (i.e. head-on situation, crossing situation, collision with stationary ships and overtaking).

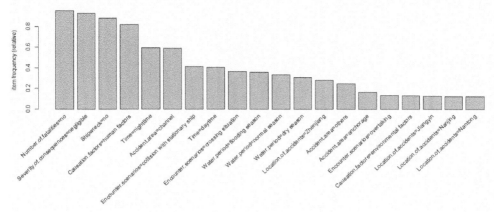

Figure 2. Frequency distribution of items for marine accidents

Table 1. Classification statistics of accident data

Variable	Description	Count	Percent
Water period (WP)	Dry season (DS)	295	30.63
	Flooding season (FS)	345	35.83
	Normal season (NS)	323	33.54
Time (T)	Daytime (DT)	390	40.50
	Nighttime (NT)	573	59.50
Location of accidents (LOA)	Zhenjiang	270	28.04
	Jiangyin	121	12.56
	Nanjing	118	12.25
	Nantong	118	12.25
	Taizhou	96	9.97
	Taicang	81	8.41
	Others	159	16.51
Accident area (AA)	Anchorage (ANCH)	158	16.41
	Channel (CHAN)	569	59.09
	Others	236	24.51
Severity of consequences (SOC)	Major	16	1.66
	Minor	52	5.40
	Negligible (NEG)	895	92.94
Shipwreck (SW)	No	847	87.95
	Yes	116	12.05
Number of fatalities (NOF)	No	916	95.12
	1-2 fatalities	38	3.95
	Over 2 fatalities	9	0.93
Causation factors (CF)	Cargo factors (FOC)	3	0.31
	Environmental factors (EF)	123	12.77
	Human factors (HF)	788	81.83
	Ship conditions (SC)	49	5.09
Encounter scenarios (ES)	Collision with stationary ship (CWSS)	399	41.43
	Crossing situation (CS)	350	36.34
	Head-on situation (HOS)	79	8.20
	Overtaking (OT)	129	13.40

From 2012 to 2016, 963 maritime accidents occurred in this waterway area. The proportions of different states for each variable are given Table 1. Specifically, the accident rate of flooding season is higher than normal season and dry season; the accident rate of nighttime is higher than daytime; the location of accident rate in Zhenjiang is obviously higher than other areas. In addition, the accident rate of channel is even higher than other water areas; the accident rate of accidents caused by human factors is around 80%, which coincides with previous statistics of accidents (Fan et al. 2017).

3.2 *Rules of ship collision accidents*

In order to derive the association rules for ship collision, the Apriror algorithm is introduced by using "arules" package in R software. In this case study, the original data of 963 marine accidents were divided into several sections and changed into CSV formats. As shown in Figure 1, the thresholds of minimum support and minimum confidence are defined as 0.15 and 0.8, respectively, and 1243 association rules are generated after using this definition.

Note that some rules share the same semantic measure or statistical measure in the extracted rule set, which is considered as redundant rules (Sahoo et al. 2015). In other words, when an association rule is the parent rule of another rule, and the parent rule has the same or lower lift, the parent rule is a redundant rule. For example, a parent rule: {location of accidents = Zhenjiang, severity of consequences = negligible, number of fatalities = 0 fatalities} => {accident area = channel}, where the lift = 1.357, and the child rule: {location of accidents = Zhenjiang, severity of consequences = negligible} => {accident area = channel}, where the lift = 1.357. This parent rule is redundant because redundant rules cannot provide more detailed and effective information, the existence or absence of the item "number of fatalities = 0 fatalities" has no practical assistance on the child rule. By removing the redundant rules, there are only 231 effective rules used for the next step.

Figure 3. Grouping matrix diagram for 231 rules

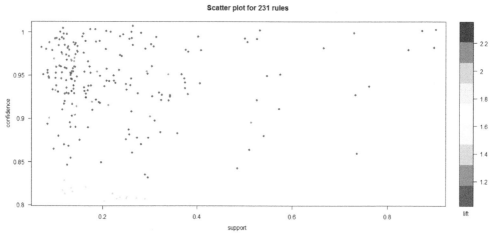

Figure 4. Scatter plot for 231 association rules of collision accidents

Afterwards, the "arulesViz" package in the R software is utilized to visualize the obtained association rules. As the factors are classified into 37 categories, the items with low support tend to have negligible influence on the rules. In order to derive the high frequency items of collision accidents, the first 20 items of frequency distribution are selected and the frequency distribution is shown in Figure 2. It can be seen that more than 95% of the accidents haven't caused casualties, over 90% of the accidents are minor accidents, 87.95% of them haven't caused shipwreck, human factors is prominent, and the occurrence of collision in Zhenjiang section is high.

The degree of lift is greater than 1 indicates that there is a positive correlation between the antecedent and the succedent (Geurts et al. 2005). By setting the threshold of the lift degree as 1, 231 positive correlation rules are obtained. The above rules are visualized in Figure 3. The default aggregate function takes the average value of the group of association rule and is represented by the color and size of the graph (Weng et al. 2016). Similar association rules are divided into one group so as to extract the general characteristics of association rules. The vertical bar represents the antecedent of the positive association rules and is clustered into 58 groups. Due to the limitation of the size of the graph, part of the antecedent is omitted, and the number of omitted items is reflected in the graph. The transverse bar represents the succedent of the positive association rules which is clustered into 6 groups. The size of the balloon indicates the support level. The larger the balloon, the higher the support degree of the association rules. Moreover, the color of the balloon indicates the lift level. The deeper the balloon color, the higher the lift degree of association rule, and more closely relationship between the antecedent and the succedent. For example, it can be seen from Figure 3 that ships in anchoring areas are more likely to collide with stationary ships.

Note that from Figure 3, the groups of high lift rules and high support rules do not coincide with each other, and there are still some hidden information needs to be further analyzed. Therefore, the scatter plot is introduced for these 231 rules and the result is shown in Figure 4. This figure can clearly show the difference and the relation among the three measures of support, lift and confidence. Each point in Figure 4 represents an association rule. The majority of support is below 0.4 while the majority of lift is between 1.0 and 1.3 with some rules more than 2.

3.3 *Analysis of high support association rules*

The higher support, the greater probability of occurrence of the item. In order to further explore the relationship between high support rules and high lift rules, the thresholds for setting high support rules and high lifting rules are 0.3 and 1.1 respectively, and only 46 rules are derived, which is shown in Figure 5. The source in Figure 5 represents the antecedent, the arrow indicates the direction of the relationship, and the end of the arrow indicates the succedent. It is discovered from this figure that the high support rules are mainly distributed in {severity of consequences = negligible}, {shipwreck = yes}, {causation factors = human factors}, {time = nighttime}, {accident area = channel}. To a certain extent, this indicates the general principle of the collision accidents in the Yangtze River section in recent years.

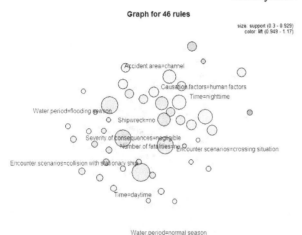

Figure 5. Graph for 46 rules of high support rules

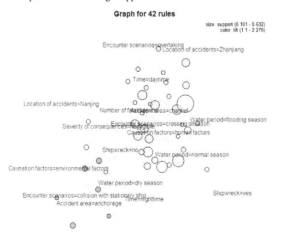

Figure 6. Graph for 42 rules of high lift rules

To further understand the rules expressed in the above graphs, the high support rules are shown in Table 2. It can be seen that the confidence of the rule {water period = flooding season} => {number of fatalities = 0 fatalities} is 0.954, which indicates that the flooding season of the Yangtze River will not have a great impact on casualties. It also shows that the characteristic of the Yangtze River collision accident is that: minor accidents occupy a high proportion, the majority of them occur in the channel and nighttime. And human factor is the main causation factor.

Table 2. High support rules for collision accidents

Antecedent	Succedent	Supp	Conf	Lift
{SOC=NEG}	{NOF=no}	0.929	1.000	1.051
{SW=no}	{NOF=no}	0.870	0.989	1.040
{SW=no, OF=no}	{SOC=NEG}	0.865	0.994	1.070
{SW=no}	{SOC=NEG}	0.865	0.983	1.058
{CF=HF}	{NOF=no}	0.773	0.944	0.993
{CF=HF}	{SOC=NEG}	0.755	0.923	0.993
{CF=HF}	{SW=no}	0.709	0.867	0.985
{SW=no, NOF=no, CF=HF}	{SOC=NEG}	0.698	0.996	1.071
{SW=no, CF=HF}	{SOC=NEG}	0.698	0.984	1.059
{T=NT}	{NOF=no}	0.567	0.953	1.002
{AA=CHAN}	{NOF=no}	0.560	0.947	0.996
{T=NT}	{SOC=NEG}	0.551	0.927	0.997
{AA=CHAN}	{SOC=NEG}	0.541	0.916	0.985
{AA=CHAN}	{CF=HF}	0.532	0.900	1.100
{T=NT}	{SW=no}	0.520	0.874	0.994
{T=NT, SW=no}	{NOF=no}	0.517	0.994	1.045
{T=NT, SW=no}	{SOC=NEG}	0.514	0.988	1.063
{AA=CHAN}	{SW=no}	0.506	0.856	0.973
{AA=CHAN, SW=no}	{NOF=no}	0.503	0.994	1.045
{T=NT}	{CF=HF}	0.501	0.841	1.028
{T=DT, NOF=no}	{SW=no}	0.353	0.919	1.045
{ES=CS}	{CF=HF}	0.348	0.957	1.170
{WP=FS}	{NOF=no}	0.342	0.954	1.003
{ES=CS}	{NOF=no}	0.335	0.923	0.970
{WP=FS,NOF=no}	{SOC=NEG}	0.334	0.979	1.053
{WP=FS}	{SOC=NEG}	0.334	0.933	1.004
{ES=CS}	{SOC=NEG}	0.324	0.891	0.959
{T=NT, AA=CHAN}	{CF=HF}	0.323	0.907	1.108
{WP=NS}	{NOF=no}	0.318	0.947	0.996
{WP=FS}	{SW=no}	0.317	0.884	1.005
{WP=FS, OF=no}	{SW=no}	0.314	0.918	1.044
{WP=FS, SOC=NEG}	{SW=no}	0.312	0.932	1.059
{WP=NS}	{SOC=NEG}	0.310	0.926	0.996
{ES=CS}	{SW=no}	0.303	0.834	0.949
{SW=no, ES=CS}	{NOF=no}	0.300	0.990	1.041

3.4 *Analysis of high lift association rules*

The lift is introduced to further discover the correlation of association rules. The higher the lift, the closer the relationship of rules is. Define the threshold value of lift as greater than 1.1, 42 rules are derived and the result is shown in Figure 6. As can be seen from Figure 6, the high lift rules are: {accident area = anchorage}, {shipwreck = yes},

{encounter scenarios = collision with stationary ships}. High support rules are: {causation factors = human factors}, {encounter scenarios = crossing situation}, {accident area = channel}. These factors are closely related to the characteristics of the collision accident in the Yangtze River, therefore, countermeasures should be taken, e.g. more attention should be paid in the anchorage as the majority of collision accident occurred in the anchorage from the rule {accident area = anchorage}.

In order to explain Figure 6, the rules with high lift values are selected and listed in Table 3. It can be seen that the high lift rule includes {encounter scenarios = collision with stationary ship}, {accident area = channel}, {causation factors = human factors}. The causes of collision accidents in the Yangtze River are closely related to these factors. According to this result, it is found that there is a tendency to collide with a stationary ship in an anchorage area, but will not cause shipwreck and the severity of consequences is negligible because quick and effective emergency response actions can be taken.

3.5 *Visualization of association rules for ship collisions*

Considering that the "arulesViz" package cannot express the relationship among all association rules, Gephi software is introduced into graphic modeling to make the rules fully visualized. The antecedent and the succedent of the rules are treated as separate nodes, therefore, the 231 rules generate 612 nodes and 1242 edges. In Figure 7, "Force Atlas" is used for layout and "Modularity Class" is used for module division. The size of nodes represents the degree of penetration, and the larger the number of connected nodes, the larger the nodes. Different colors represent different patterns, and there is a closer connection in the same patterns (Weng et al. 2016). As shown in Figure 7, the larger nodes reflect the characteristics of collision accidents more obviously, while some small nodes, which may indicate as ambiguous, is owing to the weak cause-effect relationship. The four large nodes in the figure are {severity of consequences = negligible}, {number of fatalities = 0 fatalities}, {shipwreck = yes}, {Causation factors = human factors}. This further indicates that in the studied 2012-2016 years, the Jiangsu section of the Yangtze River is mainly dominated by negligible maritime accidents, and the human factors should be drawn much attention to enhancing maritime safety.

Table 3. High lift rules for collision accidents

Antecedent	Succedent	Supp	Conf	Lift
{AA=ANCH,SW=no}	{ES=CWSS}	0.141	0.944	2.279
{AA=ANCH, SOC=NEG}	{ES=CWSS}	0.147	0.934	2.255
{AA=ANCH,NOF=no}	{ES=CWSS}	0.147	0.934	2.255
{T=NT, AA=ANCH }	{ES=CWSS}	0.101	0.933	2.251
{AA=ANCH}	{ES=CWSS}	0.152	0.924	2.230
{T=DT,CF=HF, ES=CS}	{AA=CHAN}	0.112	0.824	1.395
{T=DT,NOF=no, ES=CS}	{AA=CHAN}	0.107	0.817	1.384
{LOA=Zhenjiang, CF=HF}	{AA=CHAN}	0.196	0.815	1.379
{T=DT, SOC=NEG, ES=CS}	{AA=CHAN}	0.104	0.813	1.376
{T=DT, ES=CS}	{AA=CHAN}	0.115	0.810	1.371
{SW=no, NOF=no, CF=HF, ES=CS}	{AA=CHAN}	0.233	0.809	1.369
{LOA=Zhenjiang}	{AA=CHAN}	0.226	0.807	1.366
{SW=no, CF=HF, ES=CS}	{AA=CHAN}	0.235	0.807	1.366
{NOF=no, CF=HF, ES=CS}	{AA=CHAN}	0.258	0.805	1.363
{SW=no, NOF=no, ES=CS}	{AA=CHAN}	0.241	0.803	1.359
{NOF=no,ES=CS}	{AA=CHAN}	0.269	0.802	1.357
{SW=no, ES=CS}	{AA=CHAN}	0.243	0.801	1.356
{CF=HF, ES=CS}	{AA=CHAN}	0.278	0.800	1.354
{WP=NS, ES=CS}	{CF=HF}	0.124	0.967	1.182
{WP=FS, ES=CS}	{CF=HF}	0.110	0.964	1.178
{T=NT, SW=no,ES=CS}	{CF=HF}	0.178	0.961	1.174
{SW=no, ES=CS}	{CF=HF}	0.291	0.959	1.172
{ES=CS}	{CF=HF}	0.348	0.957	1.170
{WP=NS, T=NT, AA=CHAN}	{CF=HF}	0.115	0.933	1.140
{WP=DS, AA=CHAN, SW=no}	{CF=HF}	0.147	0.922	1.127
{WP=NS,AA=CHAN}	{CF=HF}	0.194	0.917	1.120
{WP=DS,AA=CHAN}	{CF=HF}	0.174	0.913	1.116
{CF=EF}	{ES=CWSS}	0.103	0.805	1.943
{SOC=NEG,ES=OT}	{AA=CHAN}	0.112	0.900	1.523
{NOF=no,ES=OT}	{AA=CHAN}	0.112	0.900	1.523
{CF=HF,ES=OT}	{AA=CHAN}	0.105	0.894	1.513
{ES=OT}	{AA=CHAN}	0.119	0.891	1.509
{WP=NS, CF=HF, ES=CS}	{AA=CHAN}	0.104	0.840	1.422
{T=DT, NOF=no,	{AA=CHAN}	0.104	0.833	1.410

4 CONCLUSIONS

For exploring the characteristics of ship collision accidents, the association rules method is intuitive to represent the cause-effect relationships. By establishing historical database and using association rules, it is easier to discover the causation patterns of the collision accident. For example, it is found that anchorage and berthing area are more likely to collide with stationary ships; the Zhenjiang channel has a large probability of accident occurrence, and the flooding period may easy to cause collision accidents. These findings of causation factors make the maritime safety administration, the organization in charge of maritime safety in Yangtze River, take the countermeasures to prevent the occurrence of the collision accident.

Although association rules have some advantages in the field of analyzing collision accident, there are still some shortcomings. Specifically, the evaluation standard of historical database has great influence on data results, and when dealing with different problems, the determination of the threshold of the association rules needs further analyzed.

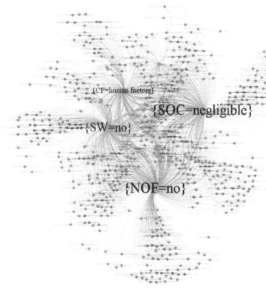

Figure 7. Network visualization of all rules

ACKNOWLEDGEMENTS

The research presented in this paper was sponsored by a grant from National Key Technologies Research & Development Program (2017YFC0804900, 2017YFC0804904), a grant from National Science Foundation of China (Grant No. 51809206), Hubei Natural Science Foundation (Grant No. 2017CBF202) and the Hong Kong Scholar Program (NO.2017XJ064).

REFERENCES:

Agrawal, R., & Srikant, R. 1994. Fast Algorithms for Mining Association Rules. *Proceedings of the 20th Very Large Data Bases (VLDB' 94)Conference*: 487-499.

Akten, N. 1999. Analysis of Shipping Casualties in the Bosphorus. *Journal of Navigation* 57(3): 345-56.

Borah, A., & Nath, B. 2018. Identifying Risk Factors for Adverse Dis eases Using Dynamic Rare Association Rule Mining. *Expert Systems with Applications*113: 233-63.

Barati, M., Bai, Q., & Liu, Q. 2017. Mining semantic association rules from RDF data. *Knowledge-Based Systems*133: 183-196.

Chai, T., Weng, J., & Deqi, X. 2017. Development of a Quantitative Risk Assessment Model for Ship Collisions in Fairways. *Safety Science* 91: 71-83.

Fan, S., Yan, X., Zhang, J., & Zhang, D. 2017. Review of Human Factors in Water Traffic Accidents. Journal of Transport Information and Safety (02): 01-08. (In Chinese)

Geurts, K., Thomas, I., & Wets, G. 2005. Understanding Spatial Concen trations of Road Accidents Using Frequent Item Sets. *Accident Analysis & Prevention* 37(4): 787-99.

Grabot, B. 2018. Rule mining in maintenance: Analysing large knowledge bases. *Computers & Industrial Engineering*, in press.

Hu, S., Fang, Q., Xia, H., & Xi, Y. 2007. Formal Safety Assessment Based On Relative Risks Model in Ship Navigation. *Reliability Engineering and System Safety* 92(3): 369-77.

Jiang, X. 2010. Statistics and Analysis of Ship Collision Accidents in Jiangsu Section of Yangtze River in Recent Five Years. China Water Transport (the second half of the month 09): 46-48. (in Chinese)

Kujala, P., Hänninen, M., Arola, T., & Ylitalo, J. 2009. Analysis of the marine traffic safety in the Gulf of Finland. *Reliability Engineering and System Safety* 94(8): 1349-1357.

Ochin, Kumar, S., & Joshi, N. 2016. Rule Power Factor: A New Interest Measure in Associative Classification. *Procedia Computer Science* 93: 12-18.

Sahoo, J., Das, A., & Goswami, A. 2015. An Efficient Approach for Min ing Association Rules From High Utility Itemsets. *Expert Systems with Applications*42(13): 5754-78.

Sedova, N., Sedov, V., & Bazhenov, R. 2018. Neural Networks and Fuzzy Sets Theory for Computer Modeling of Ship Collision Avoidance in Heavy Traffic Zone, *IEEE 2018 International Russian Automation Conference (RusAutoCon)*: 01-05.

Sii, H., Ruxton, T., & Wang, J. 2001. A Fuzzy-Logic-Based Approach to Qualitative Safety Modelling for Marine Systems. *Reliability Engineering and System Safety*73(1): 19-34.

Wang, L., & Yang, Z. 2018. Bayesian Network Modelling and Analysis of Accident Severity in Waterborne Transportation: A Case Study in China. *Reliability Engineering & System Safety*180: 277-89.

Weng, J., Zhu, J., Yan, X., & Liu Z. 2016. Investigation of Work Zone Crash Casualty Patterns Using Association Rules. *Accident Analysis & Prevention* 92: 43-52.

Weng, J., & Li, G. 2017. Exploring Shipping Accident Contributory Factors Using Association Rules. *Journal of Transportation Safety & Security*: 01-22.

Xu, C., Bao, J., Wang, C., & Liu, P. 2018. Association Rule Analysis of Factors Contributing to Extraordinarily Severe Traffic Crashes in China. *Journal of Safety Research* 67: 65-75.

Zhang, J., Teixeira, Â., Soares, CG, & Yan, X. 2018. Quantitative Assessment of Collision Risk Influence Factors in the Tianjin Port. *Safety Science*110: 363-71.

Vessel Traffic Services: Innovation, Adaptation, and Continued Relevance

G. Hanchrow
SUNY Maritime College, New York, USA

ABSTRACT: An all-encompassing aspect of voyage planning includes interfacing with vessel traffic services, whether the voyage plan includes an international transit throughout multiple VTS zones, or a domestic transit wholly limited to one VTS zone. As originally conceived, these interactive services were designed as a supplemental tool for the vessel Master by introducing real time, relevant information to assist in developing a safer transit.

The organization and function of the vessel traffic service was codified internationally within the IMO and the IALA, and nationally within port states (the USCG for example). This was in order to maintain consistency across all VTS zones for the type and level of information to be minimally provided.

Conditions within the maritime industry have changed dramatically since the inception of VTS, regarding the availability of information. Both in the qualitative technical data available, and the ability to distribute this data. Specifically, VTS was originated prior to the development of broadband satellite communication, AIS/ECDIS capability, and increased vessel automation as seen in current applications.

The intent of this paper is to look at the origin of vessel traffic services and capture the fundamental development that has transpired. Including an example the United States Coast Guard has undertaken with the NTSB to address the current effectiveness of VTS products nationwide. In addition, the paper will review a proposed level of service arisen from technological advances in data sharing. Specifically, describing the potential of "Sea Traffic Management" or "STM" as envisioned at the IMO will attempt to highlight the diverse potential of how VTS systems worldwide are adapting to changing industry conditions.

1 INTRODUCTION

The concept of managing ship movements through a shore-side radar station is generally accepted to have first appeared in the Isle of Man in 1948 with a more sophisticated and comprehensive system established in Liverpool in 1949. The integration of this new means for monitoring vessel movements with VHF radio appeared in the Port of Long Beach, California in 1951 adding a new dimension to the concept of a vessel traffic management system.

In 1956, the Netherlands established a system of radar stations for the surveillance of traffic at the port of Rotterdam. Quickly thereafter, the standard of an integrated marine radar and VHF radio system became prevalent as a means to achieve real time exchange of information between port authorities and vessels. As VTS evolved and spread in Western Europe, the commercial viability of the port was the stimulus for investing in new or expanded services.

This original effort at managing ship movements, using ship to shore collaboration and developing technology, resulted in a very detailed level of guidance internationally. IMO Resolution A.158 (ES.IV) from 1968 and IMO Resolution A.578 (14) from 1985 attempted to clarify and articulate guidelines for vessel traffic services worldwide with the goal of influencing a consistent approach to what a VTS would and would not accomplish. The current definition of a Vessel Traffic Service (VTS) can be found codified in the International Maritime Organization (IMO) Resolution A.857 (20) issued in 1997. This resolution clarified and superseded previous efforts at attempting to coordinate ship reporting systems, and recommendations for port advisory services at an international level. Additionally, the IMO developed guidance through the SOLAS Convention Chapter V effective in 2002 which provided additional direction in Regulation 12, Vessel Traffic Services. Recognizing the scale of this type of operational and procedural implementation, the IMO referred to collaborative efforts to produce effective guidance as well.

The International Association of Maritime Aids and Lighthouse Authorities (IALA) was referred to in the 1997 IMO Resolution for their reference guide detailing what should constitute a functioning VTS in accordance with the IMO intentions. While the IMO Resolution is explicit in defining VTS Requirements, this guide provided a framework on how to not only develop a functional system in a maritime domain, but how to identify a "Competent Authority" within a maritime nation to be tasked with operating and maintaining such a system. The guide also specified standards and training required to develop the operators of these systems. At the time, the personal actions of an individual VTS operator could have a tremendous effect on any given scenario requiring interpretation of vessel movement data being presented. In many cases this is still a true assessment, however the introduction and assimilation of developed technologies since the original IALA Guide was issued, has resulted in an increased level of autonomy regarding information processing. For example, a radar overlay on an electronic chart to confirm a vessel position, AIS to confirm a vessel name, and navigational details (SOG and COG). Whereas when the current guidelines on VTS were developed, these particulars were required to be monitored and confirmed without these aids by individual operators within the vessel traffic center.

According to the guidance, the VTS system should be a service implemented by a competent authority that is designed to improve the safety and efficiency of vessel traffic and the environment. The service should also have the capability to interact with the traffic in its area of operation and to respond to traffic situations that are developing. To accomplish this, the IMO promulgated the idea of four core services to be found within a VTS.

2 THE PRESENT

The first service should be that of an Information Service, (INS as detailed by the IMO Resolution). This information service is the most basic, and represents a straightforward interpretation of the VTS guidelines. Information such as Aid to Navigation (AtoN) status, and Anchorage Control are fundamental aspects of port operations and they constitute some of the basic VTS services available. Another aspect of the INS as envisioned originally is the position, identity and intentions of other vessels in the VTS area. This part of INS was exceptional in the benefit for pilots, and masters when the VTS programs came online, however currently they represent a potential for poor time management issues while piloting. This is due to the requirement within some VTS areas, especially with high traffic density, of mandatory vessel movement reporting via VHF and the ensuing return verbal communication from a VTS operator. Due to the lag in VTS requirements within some ports, this method

of vessel reporting is still required even though it is obsolete with the mandatory carriage of AIS and ECDIS. These innovations can provide an instant confirmation of the same information which can be delegated to a bridge team member for monitoring, or quickly referred to on vessels with reduced manning in pilotage waters by the pilot or master.

The second service envisioned is a Traffic Organization System (TOS as detailed by the IMO Resolution). This is the part of VTS that looks toward intervening with the advance planning of vessel movement to assist in preventing congestion, or potential unsafe navigating conditions from port activities. A detailed version of current TOS functions can be found in the USCG VTSNY guidance issued in conjunction with parameters for Ultra Large Container Vessel (ULCV) operation in the Port of New York and New Jersey. These guidelines were developed collaboratively between the USCG, Pilot Organizations, and Port Stakeholders in keeping with the way operational condition changes were intended to be addressed. However, the changes brought to New York from ULCV operations were as result in vessel design changes and operational considerations. They were not brought about by a technological advance that would possibly change the way VTS systems operate.

The samples presented below are from the Harbor Operations Committee in the Port of New York and New Jersey. They detail some of the technical information that the New York VTS has issued regarding the operation within the port of the ULCV's that began calling in 2018. As of this time, the integration of ULCV's in New York harbor has been a successful one.

Super Ultra Large Container Vessel Operations in the Port of NY/NJ

Guidelines as of June 30, 2017

- Super Ultra Large Container Vessel (SULCV)– any container vessel with a beam of 159 ft or greater
- Simulation studies of 14,000 TEU SULCV transits completed in the Port of NY/NJ.
- Working Group has established the following preliminary guidelines for transit of SULCV's above the Narrows.
- These guidelines will be amended as "hands on" experience is gained by the docking pilots.

Anchorage / Pilots / Weather

- Prior to starting SULCV inbound from Ambrose the berth must be confirmed "clear", and an anchorage spot should be confirmed available for bailout purposes.
- Two Pilots required (excluding shifting berths). One Sandy Hook Pilot and one Docking Pilot.
- SULCVs will not transit beyond the "Narrows" unless visibility in KVK is greater than 1.5 NM
- SULCVs will not transit beyond the "Narrows", if regardless of direction maximum sustained winds exceed 20 knots or maximum gusts are 25 knots or higher. Mariner's Harbor wind gauge to be used as reference station for KVK, Robbins Reef wind gauge to be used a reference station for Port Jersey Channel.

Transit Windows

- Transiting Bergen Point restricted to one hour either side of HW or LW Battery. Maximum of two SULCV's transits permitted per tide window. Vessel Arrival at "Narrows" and Sailing from berth are to be set up for two (2) hours before Battery HW or LW. A minimum three foot Under Keel Clearance (UKC) shall be required for transit from the Narrows to "Off" the berth. Due to berth controlling depth, maximum draft 49'0".
- Transiting Port Jersey Channel to/ from Global Terminal restricted to vessel arrival at Narrows and Sailing from berth with draft 47'-00" or more to be set up for one to two hours after Battery HW. Vessel arrival at Narrows and Sailing from berth with draft less than 47'-00" to be set up for one to two hours after Battery LW A minimum three foot Under Keel Clearance (UKC) shall be required for transit from the Narrows to "Off" the berth. Maximum draft 49'.00".

Terminal Obligations

- Cranes must be centered in berth and stopped until vessel is alongside and all fast.
- Cranes must be fully boomed up.
- Head lines and Stern lines cannot be more than two lines per bollard.
- There must be 100 feet between ships at adjacent berths.
- There must be a minimum of 475 feet of useable channel between berthed vessels at opposite berths along Port Elizabeth Channel.

Number of Tugs

- TWO (2) tugs to be available at the "Narrows" with a docking pilot aboard one of the tugs. This will provide for immediate tug assistance should a vessel be required to turn around, and either anchor or return to sea in the event the ship cannot continue into the KVK or Port Jersey. (Visibility restrictions or other impediment) The docking pilot and the Sandy Hook Pilot will communicate as necessary.
- For SULCVs **with a working bow thruster bound into the Kill Van Kull (KVK), TWO (2) additional** tugs will be assigned from KVK LB 9 to the Berth. For vessels bound for Port Jersey TWO (2) additional tugs will be assigned from Robbins Reef.
- For SULCVs **without a working bow thruster bound into the Kill Van Kull, five (5) tugs will be assigned** from KVK LB0 to the Berth. For vessels bound for Port Jersey a fifth tug boat **MAY be** assigned at Robbins Reef (docking pilot to assess need based upon handling characteristics of vessel and tidal conditions).
- *Note. With regards to the 3 items above, in the event that vessel handling characteristics dictate the use of special purpose escort tractor tugs tethered and operated in the "direct", "arrest" or "indirect" mode, special purpose escort tractor tugs will be substituted as requested by the docking pilot.*

Figure 1. Presentation to the NY/NJ Harbor Operations Committee June 2017. Developed and written by the DeepDraft Working Subcommittee comprising USCG Sector New York, VTSNY, Participating pilot organizations, tugboat operators and terminal operators.

Another service detailed in the IMO Resolution is the NAS, or Navigation Assistance Service. The IMO guidance simply refers to this service being important during difficult navigational or meteorological circumstances and normally at the request of the vessel or VTS as necessary. This is an area where the interaction personally in between the VTS and vessel operators has been most critical. Other than when there is an obvious and immediate concern, most interactions typically occur after a major incident such as a collision, or grounding. This may be due to a number of things including but not limited to VTS watch stander experience levels in being able to anticipate an issue from observed parameters. It could also be due to an issue onboard a vessel from a human element such as fatigue, or loss of situational awareness preventing the vessel watch stander from being able to process the available information as it is delivered. This could be passively delivered by critical information just appearing on a screen with no additional importance criteria available like an alarm. Or it could be delivered actively in the form of a radio transmission from a VTS operator. As this is recognized as a potential risk area, the IMO has developed "Message Markers" that can assist through their Standard Marine Communication Phrases initiative. The real benefit to the NAS as it appears moving forward and as increasingly complex navigation systems and automation become standard equipment, is the ability of the VTS to provide a "manual", or "analog" means to provide technical assistance to a vessel that experiences a loss of their sophisticated navigation system. This is a potential which would provide the vessel assistance in determining position, courses to steer, and bearings and ranges to danger. However, this part of a VTS interaction can reasonably be met with trepidation from vessel pilots and masters who would rightfully be guarded at the possibility of turning over any sort of positive navigation control to a shore based authority.

Lastly, although not codified within the IMO Resolution for VTS organization, is the aspect of emergency service coordination within a VTS area of operation. VTS New York has interacted with the port community beneficially in this manner over the past twenty years in high profile examples such as the unified response to coordinating port operations during the attacks of September 11th, 2001. They provided essential coordination when mobilization of port assets was necessary to perform evacuation duties and the ensuing vessel movement restrictions during the weeks that followed. Other examples worldwide of this type of VTS management can be seen in numerous extreme weather events that require close monitoring of conditions in order to issue sailing directives and terminal availability updates to vessels calling the port or VTS area during or after the event.

3 VTS IN THE UNITED STATES

Since the inception of an integrated ship to shore traffic management tool in Long Beach, the United States operated these systems in a decentralized manner. Meaning if a port community, municipality or state recognized such a system would create a benefit, then the local organization would occur to put a system in place. All the while there were some systems that worked well in high traffic density areas, and some that did not develop any recognizable "traffic management system".

On January 18, 1971, two tankers, the *Arizona Standard* and the *Oregon Standard*, collided just west of the Golden Gate Bridge and released about 800,000 gallons of bunker fuel into the San Francisco Bay in California. As a result of the accident, the "NTSB" created an investigation into the causal factors. The United States maintains an independent federal agency charged with determining probable cause of transportation accidents, promoting transportation safety, and assisting victims of transportation accidents. This agency is called the National Transportation Safety Board (NTSB).

The NTSB determined that the probable cause of the collision was the "failure or inadequacy of four different systems or subsystems, any one of which could have prevented the collision had it functioned adequately" (NTSB 1971). Among the systems listed in the report as inadequate was an experimental traffic information service. To provide this service, Coast Guard operators used land-based radar to monitor the waterway and broadcast information to participating vessels regarding the position and general movement of vessels. Vessel participation was voluntary, and Coast Guard

operators did not have the authority to direct vessel traffic.

In addition there was a succinct lack of guidance to vessel operators as to a VHF radio protocol to be followed. Such as designated frequencies for example, in the designated traffic area(s). This was a contributing factor during the collision timeline that prevented traffic management operators from communicating with the vessels involved.

The result of this investigation and the recommendations of the NTSB to the US Congress were to create legislation such as the "Ports and Waterways Safety Act of 1971"(H.R. 8140) which would provide explicit statutory authority for the USCG to establish and operate marine traffic regulation systems in the congested harbors and waterways in the US.

It wasn't until 1994 that the USCG codified this law into an actual regulation which created the current VTS system. These national regulations (33 CFR Part 161) provided the common vessel traffic management rules and policy, and additionally set the geographic descriptions of where a VTS area and mandatory vessel reporting would be required.

4 HOW IS ALL OF THIS WORKING OUT?

In 2016, the NTSB released a report that originated with the USCG requesting an analysis of their VTS operations. This study detailed 14 conclusions and 21 recommendations aimed at further reducing the risk of collisions, allisions, and groundings involving vessels operating within U.S. Coast Guard Vessel Traffic Service areas.

The study, "An Assessment of the Effectiveness of the U.S. Coast Guard Vessel Traffic Service System" (NTSB/SS-16/01), focused on the performance of the Coast Guard's VTS system, currently comprised of 12 VTS centers. The need for the study was driven by the investigation of six major commercial vessel accidents since the Coast Guard's 2009 implementation of its "Vessel Traffic Service National Standard Operating Procedures Manual." Information provided by the Coast Guard indicates collisions, allisions and groundings within VTS areas between 2010 and 2014 resulted in two fatalities, 179 injuries and more than $69 million in damage to vessels, facilities, infrastructure and the environment.

"Variance within a single safety system is itself a potential hazard and mariners traveling from one VTS to another must be able to rely on consistent Coast Guard services," said NTSB Chairman Christopher Hart. "The recommendations contained in our safety study, if acted upon, will improve the effectiveness of the VTS system throughout America's waterways. I note with appreciation the Coast Guard's openness and transparency with our investigators and the service's treatment of our study as a thorough and independent effort to improve Coast Guard operations."

There are 12 Coast Guard VTS centers that make up the VTS system, and each center is responsible for managing the traffic that operates inside its designated VTS area. Since 1994, participation in this system has been mandatory for most types of power driven commercial vessels, towing vessels, and dredge platforms while operating inside a Coast Guard VTS area.

During the years 2010 through 2014, an average of 18% of all reportable collisions, allisions, and groundings involving vessels meeting the requirements of a VTS user occurred while they were operating inside a VTS area. The most common causal factor assigned to these accidents by the Coast Guard was inattention errors by the mariners involved, which suggests that an opportunity exists for the VTS system to further reduce the risk of these types of accidents by taking a more proactive role in traffic management.

In this study, the National Transportation Safety Board (NTSB) examined the Coast Guard VTS system's ability to (1) detect and recognize traffic conflicts and other unsafe situations, (2) provide mariners with timely warning of such traffic conflicts and unsafe situations, and (3) control vessel traffic movements in the interest of safety.

Of the results the NTSB developed in their study, they reported the following. "The Coast Guard has long recognized the importance of safety risk management, but it has not been applying continuous risk assessment processes to its 12 VTS areas. Current procedures for the collection and quality control of activity and incident data do not support effective quantitative assessments of risk and safety performance within each VTS area or across the VTS system. Subsequently, these data are not regularly analyzed to identify and mitigate adverse safety trends, which has made it difficult (and in some cases impossible) to make statistically valid assessments of how well VTS centers are achieving their goal of reducing collisions, allisions, and groundings within their respective VTS areas."

The abstract of the NTSB final report can be found here: https://www.ntsb.gov/news/events/Documents/2016 _vts_BMG_Abstract.pdf

The complete NTSB final report can be found here: https://www.ntsb.gov/safety/safety-studies/Documents/SS1601.pdf

Figure 2. Vessel Traffic Service New York (VTSNY) – Photo Credit: NTSB SS1601

5 WHERE DOES VTS GO NEXT?

The technological changes in the maritime industry are well known and the current technology has found its way into all aspects of marine operations. From satellite navigation, to ECDIS and ENC with Radar overlays, to Virtual AtoN broadcast via AIS. These tools are found onboard vessels and VTS centers worldwide.

A reasonable question is where does the VTS to Vessel interface go next? Two aspects in marine operations that are gaining momentum are the integration of Electronic Data Interfaces (EDI) and the development of autonomously operated vessels.

In an effort to find solutions for increasing commercial pressures in the maritime industry, a concept called Sea Traffic Management (STM) has been introduced. A multi partner initiative developed by the Swedish Maritime Authority, has been endorsed by the European Union and the IMO. The STM Validation Project aims to introduce an extremely comprehensive system utilizing an interactive program for gaining a multitude of advantages for marine stakeholders, including carriers, shippers, port/terminal operators and regulatory agencies. According to the STM Validation Project, their stated objectives are as follows. "Sea Traffic Management connects and updates maritime users in real time with efficient information exchange. This information allows shipboard and shore based personnel to make decisions concerning the effective arrival times, route optimization, port call synchronization, and more efficient risk management".

The STM project intends to take the principles applied in the original Vessel Traffic Management Information System (VTMIS) programs, and apply current data sharing capabilities to essentially have vessels and shore facilities communicate directly with one another removing the interpretation currently seen in VTS centers, as a means to reduce operator errors as a causal factor in making inefficient traffic management decisions.

This initiative is currently being implemented with a number of established VTS centers participating in this STM program. They are doing so, by installing the technical means to communicate with participating vessels in the validation project to test the efficiency of data transfer, and the applicability of the proposed goals for the system. Participating VTS centers currently include VTS Horton in Norway, VTS Tarifa in Spain, and VTS Tallin in Estonia.

It is an exciting period of time to see various applications of new ideas and technology help in furthering the goals set forth many years ago. That is to help the maritime domain operate more safely and reliably through better communication.

In conclusion, the development of VTS systems worldwide have contributed to an increased operational confidence for vessel operators and the ports that interface with them. As is evidenced during the same time span, however, accidents can and will continue to happen from either a lack, or misinterpretation of information critical to a successful passage. It will be a worthwhile exercise to revisit the current state of VTS systems in twenty years to see how STM, and the nascent autonomous vessel developments look in comparison to today.

Specifically to see what is recognizable and to see if the challenges facing VTS systems today can be improved upon.

REFERENCES

[1] International Maritime Organization (IMO) Resolution A.857 (20) "Guidelines for Vessel Traffic Services", 1997.
[2] International Association of Marine Aids to Navigation and Lighthouse Authorities (IALA) "Vessel Traffic Services Manual", 2016.
[3] The Nautical Institute "Know Your VTS" *The Nautical Institute – The Navigator* Issue No. 18, June 2018.
[4] Hughes, Terry CAPT "When is a VTS Not a VTS?" *Royal Institute of Navigation – Journal of Navigation* Volume 62 Issue No. 3, June 2009.
[5] National Transportation Safety Board (NTSB) "An Assessment of the Effectiveness of the U.S. Coast Guard Vessel Traffic Service System" *(NTSB/SS-16/01)*, 2016.
[6] The United States National Academies of Science – Engineering and Medicine "Vessel Navigation and Traffic Services for Safe and Efficient Ports and Waterways Interim Report", 1996.
[7] Lind, Mikael; Hagg, Mikael; Haraldson, Sandra "Sea Traffic Management – Beneficial for all Maritime Stakeholders" *Transport Research Procedia E.U. 6th Transport Research Arena,* April 2016.

New Perspective for Communication Systems

A Review of NAVDAT and VDES as Upgrades of Maritime Communication Systems

S.I. Bauk

Durban University of Technology, Durban, South Africa

ABSTRACT: An actual trend in maritime communications is a need for exchange of increasingly large amounts of information among all maritime entities. For instance, vessels need to exchange data with vessel traffic service (VTS) centers; shipping companies are under huge competition pressure to adopt different weather, routing, monitoring, security and other applications; seafarers expect uninterrupted and sound connectivity with family and friends ashore, and the like. Therefore, for the purpose of broadcast communication between shore and ships – navigational data (NAVDAT) as an extension of navigational telex (NAVTEX) has been developing. Additionally, as an effort to provide route exchange services between ships and VTS centers with the ultimate goal of route optimization through cooperative decision making between the bridge team and shore operators – VHF data exchange system (VDES) as a particular upgrade of automatic identification system (AIS) has been implementing. Both NAVDAT and VDES as novel maritime communication systems have to support dynamic, flexible, scalable, service oriented, very complex and intuitive at the same time, systems-of-systems like maritime communication platform and e-Navigation for the future of navigation.

1 INTRODUCTION

The radiocommunication needs of maritime operators at sea and ashore are growing, since upcoming advanced systems used by the maritime mobile community require increased global coverage, higher data rates, video and multimedia capabilities. In fact, there is a requirement for a new digital communications infrastructure for exchange information between ship and shore, as well as between ships. Enhancing joint decision making between the ship's bridge team and the shore personnel is currently in a focus of research work and testbed experiments. Following this thread, in the paper is given a brief overview of two existing systems, i.e., NAVTEX and AIS. Also, two newly developed technological systems, which have to enable wider maritime data exchange over terrestrial and satellite links, i.e., NAVDAT and VDES are briefly described, including the perspectives they will open towards wider adoption of common maritime communication platform, or Maritime Cloud and e-Navigation around the globe.

2 NAVTEX

In the Safety Of Life At Sea (SOLAS) regulation IV/12.2 is stated: "Every ship, while at sea, shall maintain a radio watch for broad-casts of maritime safety information on the appropriate frequency or frequencies on which such information is broadcast for the area in which the ship is navigating" [1]. A radio watch can be maintained through Navigation Telex (NAVTEX), which is an international automated direct-printing system for transmitting Maritime Safety Information (MSI). MSI include navigational and meteorological warnings, meteorological forecasts and other maritime safety-related messages. These data are emitted by hydrographic or meteorological offices, Rescue Coordination Centers (RCC), etc. NAVTEX has been developed to provide a low-cost, easy and automated means of receiving MSI on board ships in coastal waters (370 km, i.e., about 200 nautical miles off shore, or maximum up to 400 nautical miles by coast station). The information transmitted may be relevant to all sizes and types of vessel and the selective messages or rejection feature enables mariners to receive MSI broadcasts that suit well their particular needs. NAVTEX provides automatic display or printout from a dedicated receiver. More precisely, international NAVTEX service means the coordinated broadcast and automatic reception on 518 kHz (MF) of maritime safety information by means of narrow-band direct-printing (NBDP) telegraphy using English language. There are two additional frequencies for transmitting MSI within this system: 490 kHz (MF) and 4209.5 kHz (HF) for safety messages in national languages [2]. NAVTEX is a part of GMDSS (Global Maritime Safety and Distress System), which is global communication system based upon automated terrestrial and satellite telecommunication sub-systems, to provide distress alerting and propagation of maritime safety information at sea. Within this context, it to be mentioned that international SafetyNET service means the coordinated broadcast and automatic reception of maritime safety information via the GMDSS Inmarsat Enhanced Group Call (EGC) system, using English language. Therefore, SafetyNET can be treated as a satellite based substitute for terrestrial NAVTEX radio services. There are 24 active NAVTEX transmission stations (A-X) and one backup station (Z), i.e., in total, 25 stations within each NAVAREA and/or METAREA. All stations transmit in the period of 10 minutes every 4 hours according to predefined timetable, and with limited transmission power in order to avoid interference [3]. NAVAREA means particular geographical sea area identified with the aim of coordinating the broadcast of navigational warnings. On the other side, METAREA means a geographical sea area established for coordinating the broadcast of marine meteorological information. It is also important to note that different NAVTEX messages are labeled with letters of English alphabet (Table 1). In other words, each class of NAVTEX message carries a different subject indicator character allowing a shipboard operator to program receiver to reject certain classes of messages that are not required (Figure 1).

However, navigational warnings, meteorological warnings, and search and rescue information cannot be rejected by operator. So, subject indicators refereeing to the letters A, B, D and L cannot be rejected by the receiver and will always be printed and/or displayed [5]. Basically, NAVTEX is maritime radiotelex terrestrial communication system, which works at both MF (518 kHz and 490 kHz) and HF (4209.5 kHz) frequency bands. It operates in the Forward Error Correction (FEC) mode. This means that is not possible to establish two-way communication between two stations, i.e., it is used only for broadcasting. As it is said before, NAVTEX belongs to the systems known as Narrow Band Digital Printing (NBDP). NBDP refers to channel bandwidth of 500 Hz, and it is based on Frequency Shift Key (FSK) modulation, while the shift between carrier frequencies is 170 Hz, and data rate is approximately 100 bit/sec [2;6]. Because of such low data rate NAVTEX might be treated as outdated, particularly within the context of developing complex e-Navigation, so called system-of-systems. In other words, NAVTEX can not transfer large amounts of data in real time for the needs of berth-to-berth navigation, including possibilities of advanced route exchange mechanism. Therefore NAVDAT system has been developing with the intention to complement (replace) NAVTEX. In the next section this newly developed system will be described in some more detail.

Table 1. NAVTEX messages labels and content.

Label	Content
A	Navigational warnings
B	Meteorological warnings
C	Ice reports
D	Search and Rescue (SAR) information, piracy and armed robbery warnings, tsunamis and other natural phenomena
E	Meteorological forecast
F	Pilot service and VTMIS (Vessel Traffic Management and Information System) messages
G	AIS messages
H	LORAN messages
I	(currently not used)
J	GNSS messages
K	Other electronic navigational aid system messages
L	Other navigational warnings, or, overflow from A
M-U	(currently not used)
V-Y	Special services for NAVTEX coordinators
Z	No messages on hand

Sources: [2;4]

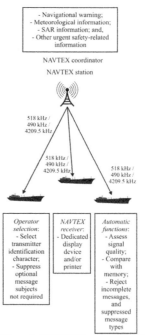

Figure 1. NAVTEX broadcasting scheme. (Source: Adapted from [7])

3 NAVDAT

Navigational Data (NAVDAT) is maritime mobile service, which operates at 500 kHz frequency band and serves for digital broadcast of safety and security information from shore-to-ship. Since it is similar in terms of its global architecture to NAVTEX, it will be coordinated by IMO in the same manner [8]. This reduces costs and makes easier the evolution from the NAVTEX to the NAVDAT. The new system uses an Orthogonal Frequency Division Multiplexing (OFDM) modulation in the 10 kHz bandwidth (i.e., 495-505 kHz) [9]. It allows two basic transmission modes: general - broadcast (i.e., to all ships) and selective - multicast, or unicats (i.e., by geographic area, by group of ships; or, to a specific ship) [10].

Besides text files, NAVDAT enables transfer of drawings, graphs, pictures, etc. These data include meteo-ocenographic information in the form of charts (isobaric, e.g.); numerical data (regular updates on the position of the eye of a tropical cyclone; tides and currents, etc.); reports showing mapped positions of ice and icebergs; warnings due to piracy dangers; SAR messages, and other maritime safety related information [11]. Data rate is considerably enlarged in comparison to NAVTEX, and it is 15-25 kbit/sec. NAVDAT provides good radio coverage of approximately 250-350 nautical miles by coast station. With this system, as it has been pointed at the World Radio Conference (WRC) in 2012: "100 years after shipwreck of Titanic, the 500 kHz is returned to the sailors". Types of NAVDAT messages are presented in Table 2.

In NAVDAT system, receiver is always on watch mode. All priority messages have to be approved by the operator. Digital Radio Mondiale (DRM) as a proven technology that provides superior coverage has been incorporated in NAVDAT system. This technology gets rid of multi-path and extends coverage from sky-wave propagation signals. It uses 16-QAM and 64-QAM modulation schemes, depending on coverage requirements, transmitter placement, antenna power and height, and the like.

It is to be said that DRM is capable of supporting Single Frequency Network (SFN). In the case of using SFN, a number of transmitters transmit on the same frequency, identical data signals. Enabling that these signals arrive within a time difference of less than guard interval, they will provide positive signal reinforcement. Consequently, the service coverage will be improved at that place compared to that obtained if there was only a single transmitter providing service [12].

Table 2. Types and content of NAVDAT messages.

Messages		
Priority	Navigation	Wide services
Navigational warning	Meteorological forecast	Update of cartography
Meteorological warning	Local meteorological information	Graph of weather evolution at certain zone
Piracy warning	Pilot information	Information for fishermen
Ice warning	Tide and current information	Harbor messages
For a specific ship (use of the MMSI)	VTS traffic	
	Aids to navigation status	
	Cartography of ice and icebergs	

Sources: Adapted from [10]

In 2012, during the World Radiocommunication Conference (WRC), 153 countries signed up to the exclusive use worldwide of the 500 kHz band for mobile maritime services. This followed the trials made onboard Pont Aven, a vessel owned by Brittany Ferry in 2010. The signals were transmitted from an experimental station near Brest in France. These signals were successfully received in south Irish Sea [11].

Also, NAVDAT tests were realized in Shanghai (China) in 2014. The tests have demonstrated that Electronic Navigational Charts (ENCs) can be corrected automatically via NAVDAT and shown in Electronic Chart Display and Information System (ECDIS). This has to reduce workload of mariners and also to improve timeless and accuracy of the charts correction. In addition, flexibility of the system is increased, e.g., NAVDAT chart update file is 78 kB in size, while the original Notice to Mariners in PDF has 2,843 kB. During these tests, MSI (images and texts relating to navigational warnings, weather forecasts, hydrographic notes, etc.) were edited in eXtensive Markup Language (XML). It has been shown, as well, that MSI in XML format can be displayed successfully on ship-borne NAVDAT terminal [13].

In accordance to the above stated it can be recognized that NAVDAT system is open to the future maritime telecommunication requirements. More precisely, it enables increasing undisturbed or smooth flow of information; provides confidentiality of some information; and, supports actual maritime mobile telecommunication systems evolution towards e-Navigation harmonized system-of-systems at sea and within maritime community in a wider context.

4 AIS

The systems for monitoring of air traffic in, and out radar range, have been firstly used in aviation. Through such systems, flight number, position and aircraft's height have been transmitted and tracked. Later, similar system was introduced for monitoring shipping traffic.

This system is called Universal Automatic Identification System (UAIS), or shortly Automatic Identification System (AIS). Initially, AIS was introduced as a collision avoidance system (as a particular supplements to radar), but to large extent has also become a system for monitoring shipping traffic.

AIS is a small radio transmitter, which broadcasts information on ship's identity (e.g., call sign, Maritime Mobile Service Identity (MMSI), position, speed, course, and the like) on maritime VHF band. Therefore, it is usually called AIS transponder.

Chapter V of IMO SOLAS Convention defines which ships and until when will be required to carry AIS. The requirements are gradually phased.

There are five different types of AIS transponders that have capacities to communicate to each other:
- Class A (for ships with requirements);
- Base stations (on shore);
- Class B (for pleasure vessels);
- AIS based Aids to Navigation; and,
- AIS-SART/-EPIRB.

AIS uses Self-organized Time-Division Multiple Access (STDMA) data format that is based on time-division multiplexing. AIS transmitter is self-organizing and transmits data every 2 sec, if the ship speed is greater than 23 knots. In the cases of lower ship speed, transmission interval is larger (3, 6, 10, or 180 sec) [14].

AIS utilizes VHF channels 87 and 88 (AIS-1 and AIS-2). So, its range is almost the same as for ordinary VHF, i.e., 20-30 nautical miles depending on radio conditions and transmitter features. If AIS data is received by a shore station, or by another ship with AIS, the ship can be identified even if there is no radar echo on the radar display.

Information that is transmitted via AIS is organized in several groups: statistical data; dynamic data; voyage-related data; and, safety-related data (Table 3).

There is no requirement that AIS stations shall be based on land. However, it has been shown that there are many benefits of developing systems that cover the coast. Due to the favorable placement of these AIS base stations, their range is quite larger than ordinary AIS transceivers, i.e., 50-60 nautical miles from the coast.

Table 3. Groups and content of AIS data.

AIS data			
Statistic	Dynamic	Voyage	Safety
Name and MMSI	Position	Static draft of the ship	This is a for of text message,
Call sign and name of ship	Time (UTC)	Type of ship	which can be either transmitted
IMO number	Course and speed over ground (COG and SOG)	Category of hazardous cargo	as a broadcast (to everyone) or selectively (to an individual
Type of ship	Heading	Destination	recipient).
Length and width of ship	Rate of turn (ROT if available)	Estimated Time of Arrival (ETA)	Selective messages are answered when
Position of GPS antenna	Navigational status (underway, at anchor, etc.)	Number of passengers (not required)	the message is received.
Height above the keel			

Sources: Adapted from [14]

4.1 AIS-based Aids to Navigation (AtoN)

AIS is specified with the idea of being usable as a navigational aid. For instance, it can be used instead of, or as a complement of racon transmitter on a light beacon. In that case, the appropriate symbol will be shown in ECDIS (Figure 2).

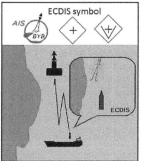

Figure 2. AIS-based buoy in ECDIS. (Source: [14, p. 2-148])

If an AtoN is out of position is marked, this will be shown in red on ECDIS. In addition to position and status message on AtoN, meteorological information can also be transmitted over the system by request. Great Britain has developed AIS AtoN with the aim to mark the most important navigation buoys [14].

4.2 AIS as a communication system

Short text messages can be sent via AIS, either to a certain MMSI within range, or as a broadcast. Primarily, AIS was not intended to be communication system. Accordingly, many experienced seafarers warn against exaggerated communication in this way instead of using speech on VHF. A specific usage of AIS as a communication tool can be for instance transmission of depth information in POB situation to the mother ship in areas of poor charting [14].

4.3 *AIS available via Internet*

Since recently several countries (e.g., Greece, Great Britain, Norway, etc.) have an open policy for spreading of AIS data via Internet free of charges [14;15]. However, it is uncertain how the different countries will deliver such services in the future, i.e., weather they will be pay-services or commercial AIS services (Figure 3).

Figure 3. AIS data distributed via Internet.
(Source: [14, p. 2-148])

4.4 *AIS-SART and AIS-EPIRB*

In accordance with GMDSS requirements, ships have to carry one or more Search And Rescue Transponder(s) (SART) based on X-band marine radar, as well as Emergency Position Indicating Radio Beacon (EPIRB). After the introduction of AIS, a revision has been made and it makes possible to use variants of AIS-SART (Figure 4) and AIS-EPIRB. These will enable persons in distress to be identified on the AIS/ECDIS display. There are also variants of Person Overboard (POB) units. AIS-SART has some advantages in comparison to radar based SART, like somewhat larger range, including ID code with GPS data, etc. The range is approximately 10 nautical miles from the ship.

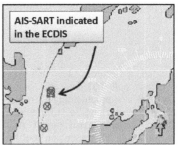

Figure 4. AIS-SART: The symbol in ECDIS.
(Source: [14, p. 2-154])

4.5 *Satellite based AIS*

With the development of AIS, the tests have been done with monitoring AIS transponders at sea from satellites. The first attempt of this kind was done by the Norwegian "student satellite" in 2004. However, the contact with the satellite after launching was unsuccessful. Later, the AISSat-1 was successfully launched and operated in 2010, and then the AISSat-2 in 2014. These are low orbit satellites (approximately, they have height of 340 km and inclination of 52°). Their range or "footprint" is approximately 1000 km and they can receive signals from about 1000 ships. This means that the receiver at the satellite will not decode all the signals. The problem with a large number of ships can either be resolved by more directional antennas and more satellites, or by introducing a dedicated transmission standard for AIS transponders [14].

Presently, ExactEarth (Canada) and Orbcomm (USA) are global leaders in satellite AIS and provide the most complete situational picture of overall vessel activity [16;17]. They provide position and identification data on over a hundred thousand unique vessels daily. In other words, they provide comprehensive and persistent geospatial intelligence for timely and accurate monitoring of vessels on a global scale. Satellite AIS data can be used not just by ships, but by naval forces, port authorities, coast guards, and other competent maritime authorities to enhance SOLAS Convention and improve maritime situational awareness. There are two methods for detecting AIS signals from the space: on-board processing (OBP) and spectrum decollision processing (SDP) [17]. Satellite AIS can be used for: pollution control; validation of ship declarations; ship routing; natural disaster relief; piracy monitoring; environmental preservation; Arctic monitoring; mitigating infectious disease; increasing vessel safety in hostile waters; search and rescue activities, and the like. Besides, AIS satellites help port authorities managing incoming traffic; provide shipping companies, insurers, and commodity owners increased visibility of shipments, and support sovereign nations surveillance initiatives [17].

4.6 *AIS development*

Upon the above stated, it becomes clear that AIS system has a far greater potential than only being anti-collision system supplementing radar. Consequently, at the World Radio Conference (WRC) in 2012 considerable expansion of AIS channels has been proposed. The extension will be supported by using VHF as a "telephone system" via coastal radio stations. The proposed channel expansion included:

- Channels 75 and 76 are reserved for long range AIS;
- Channels 27, 28, 87, and 88 are reserved for future AIS applications (including their testing); and,
- Channels 24, 25, 26, 84, 85, and 86 for use in digital transmissions, i.e., for data transmissions.

This process with the new channels can be seen as a part of actual GMDSS modernization [18].

5 VDES

Technology called VHF Data Exchange System (VDES) has been developed with the purpose of resolving problems regarding AIS channel congestion and efficiency degradation [2;19]. Namely, today AIS is being used beyond the identification and tracking ships. More precisely, besides standard AIS reports that include: static, dynamic and voyage related data; there are also general data exchange on maritime VHF band; data transmission between ships and satellites; AIS-AtoN, -SART, and -EPIRB applications; Application Specific Messages (ASM), and since recently Maritime Service Portfolio (MSP) [20]. Due to [21], some of existing analogue channels of maritime VHF band have been relocated to AIS. The use of additional channels and digital VHF system will improve AIS services, and in parallel modernize GMDSS and support development of Maritime Cloud and e-Navigation [22;23;24]. Within the context of e-Navigation, in the short term, mandatory reporting from ships might be encapsulated into ASM. On the other side, The MSP will cover a number of VTS related and other services. The scope of the MSP concept is, in fact, to align global maritime services with the need for information and communication services in a clearly defined operational area. VTS will play a central role in the coordination of MSP information (i.e., information service, traffic organization service, navigation assistance services, and the like). The list of MSP is given in Table 4.

In this context is to be pointed out that a new VDES will provide two-way communication at considerably higher data rate than previously used AIS systems. Within the VHF maritime frequency band (156.025-162.025 MHz), VDES integrates AIS with ASM and MSP to enhance smooth distribution of maritime data including extensive meteorological and traffic data. By enabling VDES to use a satellite platform, a global data exchange between ships and shore via satellites, will be enabled. It is also expected that future ship VDES transceivers will be combined with AIS into a single device. The WRC-15 recognized that the VDES satellite component is necessary to expand the system from coastal coverage to global one, and recommend that further research is to be done in order to decide on the further development of satellite VDES, during the upcoming WRC-19 [25].

Table 4. Maritime Service Portfolio (MSP).

Portfolio	Service
MSP1	Vessel Traffic Service (VTS) Information Service (IS)
MSP2	Navigation Assistance Service (NAS)
MSP3	Traffic Organization Service (TOS)
MSP4	Local Port Service (LPS)
MSP5	Maritime Safety Information Service (MSI)
MSP6	Pilotage Service
MSP7	Tugs Service
MSP8	Vessel Shore Reporting
MSP9	Telemedical Assistance Service (TMAS)
MSP10	Maritime Assistance Service (MAS)
MSP11	Nautical Chart Service
MSP12	Nautical Publication Service
MSP13	Ice Navigation Service
MSP14	Meteorological Information Service
MSP15	Real-time Hydrographic and Environmental information Service
MSP16	Search and Rescue (SAR) Service

Sources: [20, p.24]

6 CONCLUSION

From the aspect of Information and Communication Technology (ICT) management, the paper presents the evolution of NAVTEX and AIS marine radio communication systems into corresponding NAVDAT and VDES systems. Since radiocommunication is the only form of communications available at sea, it is necessary to provide a common infrastructure of establishing links between ships, and between ships and shore, including satellites as interconnection points. Seafarers and maritime entities ashore have warrant needs for more interoperable, more reliable, more functional, and more secure maritime radio communication systems. Or, if we change the perspective, issues as maritime safety and security, including protection of crew, passengers, cargo, and the marine eco-system has become extremely dependent upon efficient and reliable radiocommunication services.

In the paper described NAVDAT and VDES will undoubtedly support developing concepts of Maritime Cloud and e-Navigation, where VTS will play a key role in enabling team work between the crew and the personnel ashore. This will make a shift from ship- to shore-, or VTS-centric navigation and relieve considerably seafarers in the future. Also, this will open the whole panoply of ICT jobs ashore dealing with navigation, meteorology, hydrographs, bathymetry, cartography, geodesy, ecology, marine biology, etc. Therefore, it seems that NAVDAT and VDES are just two steeps further at a long path towards ashore assisted navigation and autonomous ships.

REFERENCES

[1] IMO Maritime Safety Committee, Sub-Committee on Navigation, Communications and Search and Rescue (NCSR). NAVTEX Manual. [Internet] 2018 [cited on 31st December 2018]. Available from: http://www.sjofartsverket.se/pages/105078/NAVTEX%20 Manual%202018.pdf

[2] Valčić, S., Mrak, Z. & Gulić M. 2016. Analysis of advantages and disadvantages of existing maritime communication systems for data exchange, *Scientific Journal of Maritime Research* 30: 28-37.

[3] ITU. 2011. *Manual for use by the Maritime Mobile and Maritime Mobile-satellite Services*, Genova.

[4] IMO. 2011. *MSC Circular 1403 – Revised NAVTEX manual*, London.

[5] Aeromarine SRT Safety Solutions. Navtex System. [Internet] 2018 [cited on 31st December 2018]. Available from:https://gmdsstesters.com/radio-survey/gmdss-radio/navtex-system.html

[6] ITU. 1995. Recommendation ITEU-R M.625-4: *Direct-printing telegraph equipment in the maritime mobile service.*

[7] MySeaTime. Rajeer, J. 20 Navtex Questions (and Answers) to Make Navtex Your Friend for Life. [Internet] 2016 [cited on 31st December 2018]. Available from: https://www.myseatime.com/blog/detail/20-navtex-question-and-answers-to-make-naxtex-your-friend-for-life

[8] Korcz, K. 2017. Some Aspects of the Modernization Plan for the GMDSS, *TransNav – the International Journal on Marine Navigation and Safety of Sea Transportation*, 11(1): 167-174.

[9] ITU-Radiocommunication Working Groups. 2014. *Working Document toward a draft new Report Maritime Radiocommunication Systems and Requirements (e-NAV14-10.3.6)*, pp. 1-68.

[10] KENTA, Pascal, O. NAVDAT: Navigational Data - System Presentation. [Internet] 2013 [cited on 1st January 2019]. Available from: https://blog.canpan.info/oprf_en/img/E58FA4E9878EE99B BBE6B0973.pdf

[11] Digital Ship. NAVDAT to be presented to IMO. [Internet] 2012 [cited on 1st January 2019]. Available from: https://thedigitalship.com/news/electronics-navigation/item/1993-navdat-to-be-presented-to-imo

[12] ITU-R.2012. Recommendation ITU-R M.2010: *Characteristics of a digital system, named Navigational Data for broadcasting maritime safety and security related information from shore-to-ship in the 500 kHz band*, pp.1-19.

[13] IMO Sub-Committee on Navigation, Communications and Search and Rescue. 2014. *Analysis of Developments in Maritime Radio Communication Systems and Technology - NAVDAT-based maritime safety related information broadcasting tests conducted in China* (submitted by China), pp. 1-7.

[14] Kjerstad, N. 2016. *Electronic and Acoustic Navigation Systems for Maritime Studies*, 1st Ed., NTNU - Norwegian University of Science and Technology, Alesund, Norway.

[15] Bauk, S. 2017. *Prilozi digitalizaciji u pomorstvu*, ELIT, Podgorica, Crna Gora.

[16] ExactEarth. An ExactEarth Technical White Paper. [Internet] 2015 [cited on 2nd January 2019]. Available from:https://cdn2.hubspot.net/hubfs/183611/Landing_Page _Documents/Satellite_AIS_White_Paper_Final-1.pdf

[17] Orbcomm. Satellite AIS. [Internet] 2001-2019 [cited on 2nd January 2019]. Available from: https://www.orbcomm.com/eu/networks/satellite-ais

[18] Bogens, K. 2017. GMDSS modernization and e-navigation spectrum needs. *ETSI Workshop "Future Evaluation of Marine Communication"*, 7-8th November, Sophia Antipolis, France, pp. 1-23.

[19] Denmark input to the IMO e-navigation CG. *An overview of the 'Maritime Cloud' – proposed information exchange infrastructure for e-navigation*, pp. 1-21.

[20] Valcic, S., Zuskin, S., Brcic, D. 2018. VHF Maritime Mobile Band – A New System to Declutter AIS Channels. *Sea Technology*, July: 24-27.

[21] ITU. 2014. Report ITU-RM.2231-1, *Use of Appendix 18 to the Radio Regulations for the marine mobile service*, Electronic Publication, Genova.

[22] Ervik, J.L. Safety and Navigation using e-Navigation Solutions. [Internet] ; [cited 3rd January 2019]. Available from:https://www.innovasjonnorge.no/globalassets/arrange menter/8---jon-leon-ervik.small.pdf

[23] Valčić, S., Pogany, T., Mrak, Z. 2018. A Model of OFDM based Maritime VHF Communication System for Data Exchange, *Polish Maritime Research* 2(98): 27-36.

[24] Bauk, S. 2018. Koncept pomorskog klauda [Maritime Cloud Concept, engl.], *Proc. of the 10th International Conference Information Technologies for e-Education (ITeO)*, 28th-29th September, Banja Luka, Republika Srpska, pp. 18-28.

[25] Bradbury, L.M. 2018. NorSat-2: Enabling advanced maritime communication with VDES, *Acta Astronautica*, in press. Available from: https://doi.org/10.1016/j.actaastro.2018.10.030

Detained Vessels Under Paris MoU: Implementation of GMDSS

R.E. Rey-Charlo, F. Piniella & J.I. Alcaide
University of Cádiz, Cádiz, Spain

ABSTRACT: The automation of communications on board ships and the approval of Global Maritime distress Safety System (GMDSS) in 1999, culminated in the disappearance of the onboard Radioelectronic officer. The GMDSS demands that captain and officers be proficient in the use of equipment and procedures but today it remains a matter of concern. The study employs as an indicator the number of deficiencies in radiocommunications detected in Port State Control inspections. For this, two studies are carried out, one generic between the years 1995-2016 and other the more detailed, based on a total of 26,795 deficiencies in ships inspected under Paris MoU. The study found that the challenges of implementing the GMDSS are related to procedures, maintenance, radio logbook and the operational capacity of the officers. This article concludes by stressing the need for more suitable training for the officers in the new technological changes and observes their excessive workload.

1 INTRODUCTION

Since the origins of commercial navigation, it has tried to prevent accidents, besides trying to increase the chances of survival in a disaster (Bueger 2014). In 1914, after the catastrophe of the Titanic, a Convention was held, from which came the Safety of Life at Sea (SOLAS), which includes rules relating to the safety of navigation and human life at sea (IMO 2014).

The SOLAS was revised and updated several times, until the International Maritime Organization (IMO) assumed responsibility for maritime transport. The IMO's fight for safety principles gave rise to a global emergency system. The adopted measure was the International Convention on Search and Rescue (SAR) and the automation of radio communications through the implementation of the Global Maritime Distress and Safety System (GMDSS). Note that the automation of communications is achieved by applying the new digital techniques to existing equipment (VHF / MF / HF) and satellite communications (INMARSAT and EPIRBS). This enables a quick response from the SAR authorities. The development of SAR plans in maritime zones has contributed significantly to maritime safety (IMO 2006).

In February 1999, the GMDSS came into force definitively, which forced passenger ships and freighters of more than 300 Gross Tonnage (GT) to comply with the requirements required by international regulations on radiocommunications (IMO 2017a).

The GMDSS is a system that provides search and rescue coordination (emergencies), maritime information for safety like meteorological, safety to navigate, etc. (security), and general communications and between ships (routine). This system requires transmission requirements, as well as the ability to receive distress alerts, both ship to land and ship to ship, for SAR operations and in disaster locations, etc. The GMDSS complies with rules 7, 8, 9, 10 and 11 of Chapter IV of the SOLAS, which allow improvise the possibilities of search and rescue operations in an accident.

The important role of GMDSS in SAR operations should be noted. The GMDSS was established to provide the SAR with efficient and well-defined communications in the activities between shore and shore, in order to carry out SAR operations effectively. Both the SAR and the GMDSS are crucial for maritime safety (IMO 2016).

With the implementation of the GMDSS, the Radioelectronic Officer, entrusted with radiocommunications, maintenance and repair of equipment on board ships, disappears (Note 1). In the writing of the GMDSS by IMO and ITU, the importance of the work performed by the Radioelectronic Officer was considered. It was in the 1988 amendments that in the SOLAS Convention it was agreed that every ship that complies with the GMDSS should have trained personnel, according to:

- International Convention on Standards of Training, Certification and Watchkeeping for Seafarers (STCW) Chapter IV Section IV / 2. (IMO 2017b);
- SOLAS 2014, chapter IV, rule 16, personal radio;
- European Committee on Radiocommunications [ERC / DEC / (99) 01], Article 9 and 10. (UIT-R 2015).

Vessels depending on the area in which they sail must comply with the maintenance requirements described in SOLAS, according to Chapter IV, Part C Rules 15.6 and 15.7.

Captain and Officers must receive the necessary training that allows them to meet the functions of the GMDSS, according to OMI Resolution A.703 (17) and A702. As agreed to by said regulations, they will obtain GMDSS General Operator Certificates or GMDSS Restricted Operator Certificate.

2 BACKGROUND

Substantial changes have taken place in maritime transport safety control in recent decades. In 2013, Yang, Wang and Li review the challenges of maritime safety analysis (Yang et al. 2013), the different approaches used to quantify the risks in maritime transportation. In 2004 and 2007, Knapp and Frances pioneered the application of econometrics in this area to more accurately quantify global PSC effectiveness (Knapp 2004; Knapp & Franses 2007a, b). In 2007, these researchers concluded that it was necessary to revise the frequency of inspections according to ship risk profile, and their recommendations were subsequently implemented by parties to the Paris MoU. The new inspection regime (NIR) took effect in 2011 (Bijwaard & Knapp 2009; Knapp & Franses 2010; Knapp & Van de Velden 2009).

In 2008, Li and Zheng studied the effectiveness of PSC and the methods adopted by regional PSC agreements to select ships for inspection (Li & Zheng 2008); their study confirmed that the enforcement of PSC is effective in terms of improving ship safety levels in maritime transport. In 2012 and 2014, more recent and novel studies include those by Bang and Li, Yin and Fan (Bang and Jang 2012; Li et al. 2014), who explored the relation between PSC inspections and a ship's involvement in accidents and incidents. Following the stream of effective PSC inspections, Özçayir (2009) studied the use of PSC in maritime industry and application of Paris MoU (Özçayir 2009). In 2014, Wu et al. have studied specific aspects of inspection (Wu et al. 2014).

At international level, maritime security has evolved from a regulatory framework based on the legislation of each country and, with interests in the maritime industry, to an international model regulated by the IMO. The Agreements, Resolutions, etc. of the IMO are fundamental tools for the regeneration and execution of the objectives that the IMO has entrusted. It is worth highlighting the work of the IMO, such as the regulations of SOLAS, STCW, MARPOL and MLC, which are frequently amended to assimilate the changes in the maritime industry (IMO 2017c).

The responsibility for compliance with the regulations, according to the United Nations Convention on the Law of the Sea in its Art.91.1 falls on the State where the ship is chartered, and therefore according to Art.94 it is this State that must take the necessary measures to guarantee safety at sea (United Nations Convention on the Law of the Sea 1982).

When flagging a ship in a country, it is obliged to comply with national requirements, be they commercial, fiscal, tax or technical. With the aim of saving costs in compliance with the regulations and requirements in terms of crew training, there is an increase in the number of open registers, and as a consequence a greater difficulty in complying with international regulations (Alcaide et al. 2016).

To reduce this difficulty and according to Rule 6 of the SOLAS, the States delegate the inspections to Rating Societies which will comply with the requirements established by the Maritime Administrations.

Another intervention measure for maritime security is the Control of the State of the Port (PSC). As pointed out and analyzed by different authors (Cariou et al 2015; Grbić et al 2015; Graziano et al. 2017; Graziano et al. 2018), consists of the inspection of foreign ships in national ports for the purpose of checking their certificates and other documentation, verifying that the conditions of the vessel, its equipment, and crew meet the requirements of the International Conventions for adequate conditions for navigation. OMI's efforts over the years have been through two different lines of action. On the one hand, the elaboration of International Agreements to be fulfilled by the Flag States and on the other, the real and effective implantation of said Conventions by the States that ratify them.

The Maritime Administrations in Europe are under the protection of the Memorandum of Paris (MoU). The MoU is based on the international regulations of the seven most important agreements affecting maritime navigation. Therefore, the monitoring of maritime communications, as part of Chapter IV of the SOLAS, are the responsibility not only of the State of the flag but also and more effectively of the PSC inspections.

This study is based on the survey of communication deficiencies provided by the MoU of

Paris, which includes the maritime zone of Europe and Canada (Atlantic coast). This Memorandum includes 27 maritime authorities that cover the entire European area and the North Atlantic.

Study 1: a generic analysis between the years 1995-2016, period in which the presence of the Radioelectronic Officer complied with the requirements of the traditional relief system until the final implementation of the GMDSS through the period of coexistence of both systems.

Study 2: a more detailed analysis, based on 26,795 deficiencies found in communications in the period from 2005 to 2016 and classified by type, age, State of registration, etc.

3 METHODOLOGICAL ASPECTS

The analysis proposed in our document is based on the use of the database of PSC inspections carried out under the framework of the Paris MoU. In the period from 2005 to 2016, as shown by the sources consulted, a total of 247,883 inspections of the Paris MoU are available, of which it has been analyzed 26,795 registry entries in maritime radiocommunications.

In order to present the incidents of radio-communications, based on the inspections of the PSC, it is a useful source since there is talk of a very important part for the SAR and therefore for maritime safety.

The coding used by the Paris MoU in deficiencies is shown in Table 1. In the case of non-compliance with the regulations, those of group 05 correspond with deficiencies in the safety of the equipment or non-compliance with the procedures of maritime radiocommunications.

Table 1. List of Paris MoU deficiency codes. Source: Paris MoU.

Code	Deficiency
01	Certificates & Documentation
02	Structural condition
03	Water/Weathertight condition
04	Emergency Systems
05	**Radio communication**
06	Cargo operations including equipment
07	Fire safety
08	Alarms
09	Working and Living Conditions
10	Safety of Navigation
11	Life saving appliances
12	Dangerous Goods
13	Propulsion and auxiliary machinery
14	Pollution Prevention
15	ISM
16	ISPS
17	MLC,2006
99	Other

The codification established by the Paris MoU allows focusing on deficiencies in maritime communications, in order to obtain a global view of the ships and registers most affected and the analysis of the evolution of the values of inspections of the countries that make up the various MoUs.

Within the deficiency group 05, it can be differentiated the different types of breaches of the safety standards in maritime radiocommunications, as established in Table 2 (Note2).

Table 2. Type of deficiencies code. Sources: Paris MoU.

Code	Type of Deficiencies
5101	Distress messages: obligations and procedures.
5102	Functional requirements
5103	Main installation
5104	MF radio installation
5105	MF/HF radio installation
5106	INMARSAT ship earth station
5107	Maintenance/duplication of equipment
5108	Performance standards for radio equipment
5109	VHF radio installation
5110	Facilities for reception of marine safety information
5111	Satellite EPIRB 406MHz/1,6GHz
5112	VHF EPIRB
5113	SART/AIS-SART
5114	Reserve source of energy
5115	Radio log (diary)
5116	Operation/maintenance
5118	Operation of GMDSS equipment
5199	Other (radio communication)

The database of the Paris MoU consulted, has the IMO number, name of the vessel, flag, type of vessel, keel keying, tonnage, country code, name of the port, date of inspection, type of inspection, deficiency, deficiency code action and, whether or not the inspected vessel is detained.

4 RESULTS AND DISCUSSION

Seafarers are at constant risk and at any time they may need immediate assistance, therefore, contracting governments committed either individually or in cooperation with other governments to provide appropriate facilities that comply with the GMDSS. in terms of safety according to SOLAS, Chapter IV Rule 5 both on land and on the ship.

The search and rescue at sea is a duty, so that assistance has always been provided to ships and people who need it. Therefore, a deficiency in the GMDSS installation in a ship is something to be taken into account and study, since any deficiency directly affects SAR operations and therefore human life and environmental pollution.

As a starting point the issue has been addressed in the PSC inspections, where radiocommunication deficiencies have been related to different parameters such as ship age, type, registry, inspection port, country, etc. References that are adopted to classify ships according to their level of risk.

The information obtained from the Paris MoU's annual reports shows the number of deficiencies in communications without relating them to any parameter. Figure 1, presents the absolute values of deficiencies. Figure 2, shows on a relative manner in the total of all the deficiencies found in the vessels whose inspections were carried out in that period.

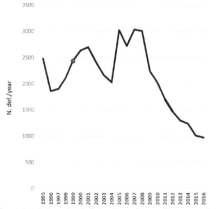

Figure 1. Nº of deficiencies in radiocommunications 1995/2016. Source: Authors' own elaboration based on Annual Report of Paris MoU.

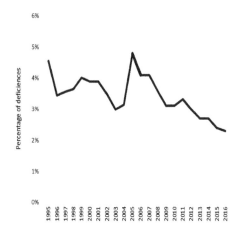

Figure 2. Percentage of deficiencies in communications of the total number of deficiencies detected on ships (Paris MoU). Source: Authors' own elaboration based on Annual Report of Paris MoU.

In both cases it manifests how during the period of entry into force of the GMDSS, until its full implementation in 1999 (with the coexistence of the two communication systems, the traditional system operating manually by the Radioelectronic Officer and the GMDSS by the Officer on watch). It can be highlighted a slight increase in the deficiencies in

the initial implementation period. On the other hand, the trend of the data does not indicate a greater number of deficiencies in communications in the PSC inspections of the Paris MoU, but a downward trend of them, and as a result, it can be established a firm consolidation of the GMDSS and its operators.

The second analysis of greater depth for the period from 2005 to 2016, it was being carried out through a segmentation of the data from different variables. The first of these has been to classification (within the code group 05) of each one of the different deficiencies, as indicated in Table 2. The results are shown in Figure 3.

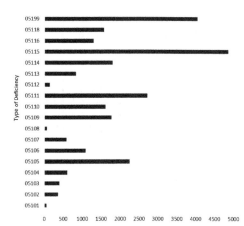

Figure 3. Types of deficiencies in communications (2005/2016). Source: Authors' own elaboration based on Annual Report of Paris MoU.

From the reading of the data in figure 3, it is observed that the most frequent deficiencies correspond to the GMDSS Radio logbook, all the ships inspected with this deficiency are due to the absence of the diary or that their records are incorrect or incomplete. In the same way it is observed that the monthly equipment tests are not carried out or are not reflected.

Another deficiency is with respect to the on-board communication station. The breach of the regulations based on the regulatory equipment according to the navigation area and its maintenance should be noted. It has been observed from broken, damaged, unconnected antennas, the antenna cable cut, damaged equipment, non-existent power sources and inadequate lighting. And one last shortcoming to highlight, all related to the documentation that must be up to date but was not updated: expired certificates, maintenance contracts not renewed, regulations not in force, the license of the incorrect station, the room of battery without signaling

Based on the results, in the table 3 is observed that the number of these deficiencies decreases as the years pass in the studied period, going from a percentage of 4.80% to reaching a figure of 2.30% in the last year (2016). This can be understood as a greater awareness, on the part of the shipping companies, in matters of maritime safety pressurised by the demands of the PSC inspections and their subsequent consequences.

Table 3. Deficiencies. Source: Authors´ own elaboration based on Annual Report of Paris MoU.

	Def. Communication	Total def.	Def %
2005	3027	62434	4,80%
2006	2724	66142	4,10%
2007	3040	74713	4,10%
2008	3009	83751	3,59%
2009	2243	71911	3,12%
2010	2200	64698	3,38%
2011	1704	50738	3,33%
2012	1476	49261	3,00%
2013	1301	49074	2,65%
2014	1242	46224	2,70%
2015	1015	41777	2,40%
2016	976	41857	2,30%
TOTAL	23957	702580	3,41%

The second most significant variable in the study was the age of the vessels, which influences the increase in the number of this type of deficiencies, since age is closely linked to the deterioration of the equipment, therefore, to deficiencies that would entail Possible detention. In Figure 4, the deficiencies of vessels over 20 years of age are especially highlighted, which determines the greatest number of deficiencies that lead to the detention of the vessel.

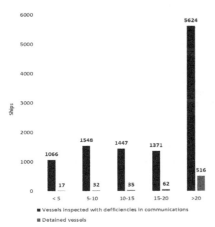

Figure 4. Deficiencies in communications reported and detained vessels according to the age of the vessels (2005/2016). Source: Authors´ own elaboration based on Annual Report of Paris MoU.

A third variable would be the one related to the flag. It is observed that the countries with the highest number of deficiencies coincide with the countries that have consolidated open registries in the world

fleet. The term "Other" represents the unification of those countries with a low number of registered vessels as shown in Figure 5.

The resolution of detentions is not exactly proportional to the size of the record. The main registries have a high standard in compliance with the regulations, not being the case in third-party registrations, where the economic or commercial benefit reigns over compliance with international regulations or agreements.

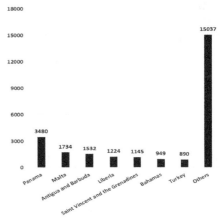

Figure 5. Deficiencies in communications according to the registry of the vessel. Source: Authors´ own elaboration based on Annual Report of Paris MoU.

The fourth variable to take into account would be the type of ship. In Figure 6, the results have been segmented by the ship typology, finding the ship with the greatest deficiencies in communications to be the general cargo ship followed by the bulk carriers. The progressive reduction of these deficiencies throughout the period analyzed is also clearly observed.

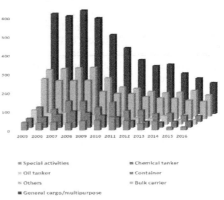

Figure 6. Deficiencies in communications according to the type of the vessels. Source: Authors´ own elaboration based on Annual Report of Paris MoU.

It is observed that these deficiencies lead to some actions, but that in most of the cases they are corrected and do not lead to the planned detention, as shown in Figure 7.

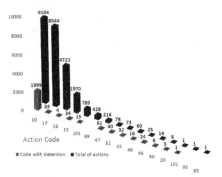

Figure 7. Deficiencies in communications according to the action code. Source: Authors´ own elaboration based on Annual Report of Paris MoU.

As in the previous analyzes, this type of deficiencies also shows a heterogeneity in the results when it is analyzed them State by State, as can be seen in Table 4 (Anexo). There are countries in which the percentage of communication deficiencies exceeds 10% and others in which it can be reached the general ratios of Table 3. The same happens if it is analyzed the inspections that involve detentions where this disparity of results is also appreciated.

5 CONCLUSION

Operators, equipment, procedures and the entire radiocommunication system, as a whole, constitute a fundamental pillar for the safety of human life at sea, as well as for the safety of navigation. The results obtained show a gradual decrease in the number of deficiencies in the period under study. However, a sector vulnerable to major human and environmental catastrophe should not depend on the possible operation of the equipment or the operator's ability to respond to marine casualties. Therefore, it would be important to reduce the magnitude about the deficiencies detected in radiocommunications.

The deficiencies in radiocommunications detected in the ports of the Paris MoU can be taken as a good indicator of the degree of implementation, monitoring, application and compliance with regulations, in addition to the competence and the degree of training of Captains and officers. On the other hand, the main factors that condition radiocommunication deficiencies (ship type, age, flag, etc.) can be of great value when defining the health of a ship registry.

The speed of the port operations and the operative itself, together with the reduction in the number of crew and the long embarkation periods, can subject the crews to an overload and work stress that can have a negative effect on the management and compliance of tasks, both maintenance and commissioning of appropriate procedures in radiocommunications.

BIBLIOGRAPHY

Alcaide, J.I; Piniella, F; Rodriguez-Díaz, E.; 2016. The "Mirror Flags": Ship registration in globalised ship breaking industry. Transportation Research Part D: Transport and Environment vol.48, pages 378-392.

Bang, H.-S., Jang, D.-J., 2012. Recent Developments in Regional Memorandums of Understanding on Port State Control. Ocean Dev. Int. Law vol.43, pages 170–187.

Bijwaard, G.E., Knapp, S., 2009. Analysis of ship life cycles— The impact of economic cycles and ship inspections. Mar. Policy vol.33, pages 350–369.

Bueger C.; 2014. What is maritime secury?. Marine Policy Volumen 53, pages 159-164.

Cariou P.; Wolff F.C.; 2015. Identifying substandard vessels through Port State Control inspections: A new methodology for Concentrated Inspection Campaigns. Marine Policy vol.60, pages 27-39.

Graziano A.; Schröder-Hinrichs,J.U; Ölcer, A. 2017. After 40 years of regional and coordinated ship safety inspection: Destination reached or new point of departure? Ocean Engineering vol.143, pages 217-226.

Graziano A; Cariou P.; Wolff F.C; Schröder-Hinrichs, J.U. 2018. Port state control inspections in the European Union: Do inspector´s number and background matter? Marine Policy vol.88, pag 230-241.

Grbić, L; Ivanišević,D; Čulin, J. 2015. Detainable Maritime Labour Convention 2006-related deficiencis found by Paris MoU authorities. Multidisciplinary Scientific Journal of Maritime Research vol.29 nº1 pages 52-57.

IMO, 2006. Search and Rescue (SAR), 2006 Edition. IMO Pub.

IMO, 2014. Safety of Life at Sea (SOLAS) Consolidated Edition, 2014. IMO Pub.

IMO, 2016. The International Aeronautical and Maritime Search and Rescue (IAMSAR MANUAL), Vol I, II, III, 2016 Edition.

IMO, 2017a. Global Maritime Distress Safety System (GMDSS Manual), 2017 Edition. IMO Pub.

IMO, 2017b. International Convention on Standards of Training, Certification and Watchkeeping for Seafarers (STCW Fishing 95), 1996 Edition, STCW inc. 2010 Manila Amendments, 2017 Edition. IMO Pub.

IMO, 2017c. International Convention for the Prevention of Pollution by ship (MARPOL) Consolidated Edition 2017. IMO Pub.

Knapp, S., 2004. Analysis of the Maritime Safety Regime: Risk Improvement Possibilities for the Port State Control Target Factor (Paris MoU). Msc. Thesis in Maritime Economics and Logistics. Rotterdam: Erasmus Universiteit.

Knapp, S., Franses, P.H., 2007a. Econometric analysis on the effect of Port State Control inspections on the probability of casualty. Can targeting of substandard ships for inspections be improved? Maritime Policy vol.31, issue 4, pages 550–563.

Knapp, S., Franses, P.H., 2007b. A global view on port state control: econometric analysis of the differences across port state control regimes. Maritime Policy and Management vol.34, issue 5, pages 453–482.

Knapp, S., Franses, P.H., 2010. Comprehensive review of the maritime safety regimes: Present status and recommendations for improvements. Transp. Rev. 30, pages 241–270.

Paris MoU. Annual Report on Port State Control, 1995–2016.

UIT-R, 2015. TU Radiocommunication Sector. Recommendations 2015.

UNCLOS, 1982. United Nations Convention on the Law of the Sea December 1982.

NOTE 1:

The article talks about the disappearance of the radioelectonic officer onboard, but it does not refer to a radical disappearance, but its presence is not obligatory. Therefore, some shipowers prefer an officer to be responsible for radio communications before having a radioelectronic officer. The presence of a radioelectronic officer implies an increase in costs.

However, it should be taken into account that the passenger ships enlist these radioelectronic officers, so this training is still given with its subsequent certification.

NOTE 2:

The MoU of Paris offers a table with the codes of deficiencies in radiocommunications. In this table, the code 5111 refers to the EPIRB 406MHz and the EPIRB 1.6GHz. It should be taken into account that, according to the MSC79-23, the EPIRB 1,6GHz was no longer in service in 2006.

ANNEX

Table 4. Radiocommunication deficiencies in relation of the total number of deficiencies according Port States of MoU (period 2005/2016). Source: Authors' own elaboration based on Annual Report of Paris MoU.

MoU Port State	Inspection			Inspections with detentions		
	Total	Communication	%	Total	Communication	%
BE	7752	439	5,66%	486	29	5,97%
BG	3357	433	12,90%	193	15	7,77%
CA	4662	219	4,70%	315	0	0,00%
HR	2414	112	4,64%	185	9	4,86%
CY	1209	117	9,68%	244	26	10,66%
DK	2492	109	4,37%	128	4	3,13%
EE	933	54	5,79%	24	1	4,17%
FI	1101	42	3,81%	39	0	0,00%
FR	9906	374	3,78%	643	31	4,82%
DE	8504	684	8,04%	517	10	1,93%
GR	6717	513	7,64%	639	65	10,17%
IS	347	3	0,86%	24	0	0,00%
IE	2216	116	5,23%	233	4	1,72%
IT	12701	774	6,09%	1839	128	6,96%
LV	2071	73	3,52%	25	0	0,00%
LT	1901	127	6,68%	46	2	4,35%
MT	1533	131	8,55%	129	21	16,28%
NL	9538	831	8,71%	535	23	4,30%
NO	2495	77	3,09%	139	4	2,88%
PL	4472	524	11,72%	259	20	7,72%
PT	3828	426	11,13%	305	23	7,54%
RO	5077	525	10,34%	328	25	7,62%
RU	9735	1307	13,43%	598	39	6,52%
SI	1326	50	3,77%	324	7	2,16%
ES	14751	1446	9,80%	1366	136	9,96%
SE	2409	33	1,37%	79	2	2,53%
GB	13516	1486	10,99%	743	40	5,38%

Challenges in Safety of Sea Transportation

Analysis of Accidents During Maritime Transportation of Cargo Fumigated by Phosphine: Causes, Consequences, Prevention

E. Belobrov, V. Torskiy & G. Rangayeva
Odessa National Maritime Academy, Odessa, Ukraine

ABSTRACT: The article presents an analysis of long-term investigation of marine accidents associated with the constantly growing export of grain fumigated with phosphine in Ukraine (more than 44 million tons in 2018) transported on ships of the "River-Sea" type and bulker fleet, and accidental leaks of toxic fumigation gases phosphine accompanied by mass cases (54) of poisoning with the death (9) of crew members and huge financial costs. In accordance with the requirements of Resolution IMO MSC 255 (84) there are comparative materials of the investigation on the causes of seafarers' deaths on ships belonging to various countries while transporting fumigated goods.

Among the main causes of accidents and poisoning of seafarers are: insufficient or lack of professional training and competence of officers on ships, cargo fumigators and port supervision, poor professional supervision of the fumigation activities of "chance" fumigation teams operating in ports without the required IMO documents, as well as failure in regulating the legislative, administrative and legal responsibility of managers of various levels.

Practical conclusions have been made and a set of preventive measures for accidents, acute poisoning of seafarers and the exclusion of unreasonable financial losses are given.

1 INTRODUCTION

Labor of seamen of the merchant fleet has always been organically linked with specific changing long-haul conditions, when crew members, according to research, are at constant risk of exposure to harmful and dangerous factors of the ship's production environment, these are: physical, chemical, biological, psycho physiological and social components which often cause accidents. , injuries, poisonings, and diseases (ILO, 2014) The number of risks of occupational poisoning which causes seafarers' death is significant and it increases many times due to growth of transportation of dangerous and quarantine cargoes in the holds of vessels which are subjected to fumigation and are in an environment of extremely toxic and poisonous pesticides, 4.3 hazard class 6.1 of IMO IMSBC CODE.

This is especially evident in emergency situations and during uncontrolled leaks of phosphine fumigation gas from holds (Nielsen, 1996; Belobrov, 2007; Golikov & Repetey, 2013). Many years of participation in the work of the maritime investigative commissions of Ukraine in the course of investigation of accidents involving phosphine containing and fumigated goods on ships and mass fatal poisoning by phosphine of seafarers allowed not only to identify obvious, but also to establish the hidden sides of the insolvency of legislative, legal, organizational, regulation of the safety of marine fumigation of cargo and the development of accident prevention measures (Belobrov & Repetey, 2006; Belobrov, 2017) and the safety of transport of fumigated goods, including ice stems conditions (Torskiy, Nezavitin, 2012).

However, at the moment there are not only documents, but also methodological approaches to organizing the monitoring of the fulfillment of the unified requirements (Recommendations IMO MSC, 2008; MLC-2006) for the maritime fumigation safety, the prevention of accidents, occupational safety and health of seafarers to the participants of the marine fumigation business (ILO, 2014). This is especially true of marine fumigators, boat masters and PSC inspectors.

The purpose of this study is based on the results of scientific and practical work on the analysis of the causes of the risk of accidents during the transportation of fumigated goods on ships, to develop a set of measures for implementation on commercial ships to prevent accidents and acute poisoning of seafarers for preserving their life and health.

2 MATERIAL AND METHODS

2.1 *Research materials*

The research was conducted since 2006 to 2016 and included an analysis of the work-related accidents with seafarers which are always associated with an accidental leakage of toxic fumigation phosphine gas from the cargo holds, which resulted in massive fatalities with seafarers on board. Fumigated cargo incidents occurred in 4 Ukrainian and 4 foreign ports on 11 old ships under a foreign flag of mixed river-sea voyages and new grain-carriers of the bulker fleet: "Coaster", "Handysize" and "Panamax" on port raids and during the voyage. All merchant ships varied in carrying capacity from 2500 to 60,000 gross registered tons, the number of holds 2-7, the number of crew members 10-42 people. We studied the data investigation into the causes of accidents on vessels that were occupied transporting regulated cereal grains, legumes, oil and feed fumigated goods, and were in flight subjected to decontamination phosphine (PH3) which is thicker cargo holds formed from tablets explosive fumigation preparations "Aluminum Phosphide "(AlP) or" Magnesium Phosphide "(Mg3P2) - 4.3 and 6.1 pesticides of hazard classes according to IMO IMDG CODE, in quantities from 2 to 700 kg per vessel.

Table 1. Distribution of the causes of accidents in the transport of fumigated goods

Country	Genus of emergency fumigated and dangerous goods	Number of reports, %
Ukraine	Grain and feed oils (4), ferrosilicon (3)	7 (63,6)
Poland	Shea Nuts (1)	1 (9.1)
Norway	Cereals (2)	2 (18.2)
Russia	Cereals (1)	1 (9.1)
Total		11 (100)

The research work included the phased collection of information, the search for literature data and a comparative analysis of the materials obtained, the study of the causes of accidents of professional cases of mass phosphine poisoning of sailors during transportation of fumigated quarantine goods. The studies were based on the "Reports of the State Commission for Investigation and Analysis of Accidental Maritime Incidents of the Ministry of Infrastructure of Ukraine", "Reports of the Commissions of Captain Services of the Seaports of Ukraine for Investigation of Accidents in Maritime Accidents", "Report of the State Marine Accident Investigation Commission of the Poland", etc. The data in Table 1 indicates that the following documents were used as the main materials of the studies: the investigative commissions of Ukraine (63.6%), other materials of the commission for the investigation of emergency marine accidents with

phosphine were used as comparative Poland, Norway, Russia, etc. (36.46%).

2.2 *Methods*

Based on the data of the authors' publications (2006-2017), materials from their own researches on the investigation of incidents and reports of marine accidents during the transportation of phosphine fumigated quarantine cargo, analysis of reports of state maritime commissions under the IMO Resolution MSC.255 (84) using: methods: regulatory - search, comparative expert evaluation, statistical processing of materials. The variable values were divided into the following groups: 1) a group of vessels, fumigation cargo and toxic fumigants, including type, flag, dimensions, cargo capacity, number of holds, number of crew members; the name of fumigated grain, fodder, oilseeds and agricultural cargoes , Shea nuts ; pesticide name in tablets, trade mark, dose and total amount of pesticide in the holds, method of distribution of the pesticide in the load, tablets ; 2) a group of training and special professional training of navigators, navigation personnel and fumigators of marine fumigation of cargo; 3) a group of variables characterizing the causes and severity of the accident, the classification and severity of phosphine poisoning with the death of sailors on a ship, the type and effectiveness of emergency medical care during the voyage, consequences of poisoning sailors. the provision of special first-aid kits for phosphine poisoning on the ship.

3 RESULTS AND CONCLUSION

As can be seen from Table 2, all the ships on which an accidental phosphine leak with the poisoning of seamen occurred differed significantly in type, number of holds, cargo capacity, kind of transported regulated cargo and a total amount of poisonous gums in the ship hold before the voyage. Despite the apparent repeated, more than 70 times, excess of chemicals delivered on the ship between River-Sea vessels and Panamax bulk carriers (10 and 700 kg!), The dose of toxic drugs determines the critical risk of dangerous contact and poisoning of seafarers, varying from 1.0 to 10 grams per 1 m3. The studies did not reveal a logical relationship between the characteristics of the vessels under study, the type of cargo being transferred, the use of toxic chemicals during marine fumigation, and the number of cases of fatal poisoning of seafarers with phosphine.

Table 2. Characteristics of types of vessels in which the accident occurred with phosphine poisoning crew

Ship flag	Type	Amount	Kind and Qty of cargo, tons	T Qty of fumigant, kg	Crew
Belize	River-Sea	2	Bran granules 2 450	10-15	9/9/2 *
Panama	Panamax	9	Corn 51 000	250 -700	28/7/2
Marshall Islands	Handysize	6	Barley 42 000	150-450	42/12/0
Russia	River-Sea	3	Wheat 5 000	15-20	12/12/1
Malta	Coaster	4	Shea Nuts 7,250	22	17/15/2
Liberia	River-Sea	3	Corn 4 250	25-30	12/11/2

Designation: * - Number of crew / total poisoning / death

Table 3. The causes of accidents, poisoning of sailors and solutions of the safety during the transportation of fumigated goods

Main causes of accidents	Practical problem-solving of accidents	Number of respondents	Feedback Positive %	Feedback Negative %
Lack of learning on marine fumigation of cargo fumigators	Compulsory education on courses IMO Marine cargo fumigation	86	86 (100)	-
Lack of learning on marine fumigation of shipmaster cargo	Compulsory education on courses IMO Marine cargo fumigation	82	76 (92,7)	6 (7,3)
Lack of practice for monitoring certificates of competence of fumigators arriving on a fumigation vessel	Mandatory requirement checking certificates training courses on marine fumigation of cargo	86	86 (100)	-
Lack of classification of grain cargo, transported in phosphine environment to 6.1 class	Recommended for the period of transportation of regulated cargo in the environment of poisonous gases, class 6.1 class	78	71 (91,0)	7 (9,0)
Lack of reliable ways to control the tightness of holds with fumigated goods before the voyage	Obligatory use of test - phosphine detector gas when checking the reliability of sealing of holds with fumigated goods before the voyage	85	81 (95,3)	4 (4,7)
Lack of practice reliability of sealing of holds with fumigated goods before the voyage	The establishment of mandatory the practice of monitoring the reliability of the sealing of holds with fumigated cargoes of the absence of phosphine in the premises of the superstructure before a voyage	76	66 (86,8)	10 (13,2)
Lack of ways sealing and one-time drainage of condensate drainage pipes during transportation fumigated goods	It is recommended to use a device that allows sealing the drainage valve and pipes of the coaming holds and simultaneously draining the condensate	80	68 (85,0)	12 (15)
Lack of practice A pre-trip check of air of the ship superstructure for the presence of phosphine gas by fumigators	Recommended practice for mandatory fumigators to check the presence of poison in the superstructure air after the end of fumigation work	86	86 (100)	-
The absence of a special first-aid kit for a voyage to assist with phosphine poisoning on board of the vessel	Required availability of the special equipment for the voyage in case of phosphine poisoning during the transportation of fumigated goods	84	80 (95,3)	4 (4,7)

However, it should be noted that the number of accidents and deaths of crew members caused by phosphine poisoning during sea transportation of fumigated goods is most often determined in the group of old technically unprepared river-sea vessels, in which the total amount of pesticides used is not that high. The analysis of the cases of treatment of injured sailors who sought for medical advice, which were engaged in the transport of phosphine fumigated grain, fodder and agricultural (Shea nuts) showed that, despite the fact that the cargo carried in the holds in the environment of poisonous phosphine gas on board, 5 sailors (45.4% of cases) made the wrong diagnosis and treated as food poisoning (food toxic infection), because the leading symptoms of phosphine poisoning are nausea , vomiting , diarrhea . For example,

according to the data of SMAIC-2015, when 17 sailors were poisoned with phosphine during the transportation of fumigated nuts, Shea nuts were originally diagnosed with food poisoning by 3 sailors of the ship's naval officers. The same diagnosis by radio medical care was confirmed by the doctor-consultant of emergency radio-medical care, as well as by the doctor of the hospital of another port of emergency call of the vessel. The true diagnosis of phosphine poisoning was established only on day 5 — during this period, 12 more people turned up with similar symptoms and 2 seamen died. Upon further consideration of Table 2 - in one case (9.6%) provided assistance for acute respiratory disease and bronchitis, two sailors (18.1%) were not diagnosed at all and provided assistance as a poisoning with an unknown gas, and only 3 Seafarers (27.8% of cases) were diagnosed correctly - phosphine poisoning with fumigation gas. The findings of unqualified home care and medical assistance to victims of accidents with fumigated cargo and seamen poisoning with phosphine indicate insufficient training and lack of competence of ship officers and employees of coastal medical institutions in providing medical assistance and other areas of knowledge and experience in marine cargo fumigation (Belobrov E. & Torskiy. 2014).

The research results presented in Table 3 show not only the volume of large representative material devoted to the study of the causes of accidents and acute poisoning with poisonous fumigation gas during transportation of fumigated goods on ships but also encourages the development of more effective preventive measures to prevent accidents, poisoning, preserving of life and health of workers in fumigation and maritime occupations. The answers to the causes of accidents involving the carriage of fumigated goods by the carriage prompted: first, to substantiate scientifically developing practical solutions of the problem, second, to study the positive and negative feedback from navigators, marine fumigation specialists, port inspectors, and preventive medicine scientists (Golikov, V. &, Repetey V. 2013). Table 3, as the main problem of accidents and casualty during the transportation of fumigated goods, distinct lack of the training of navigators, their crews and fumigators according to international rules and safety standards for maritime fumigation of goods. Therefore, all 86 respondents, in 100% of cases positively reacted to the recommendations of mandatory advanced training in IMO courses with the receipt of Certificates, confirmed by the competence of navigators and fumigators on marine cargo fumigation as an important solution to preventing accidents and casualties, poisoning and death of people on the ships (SOLAS, IMO MSC 1264, IMFO).

One of the significant causes of accidents with cargoes, transported in the holds of poisonous phosphine gas, is the lack of knowledge in the practice of marine fumigation of cargoes of the hidden processes of accidental leakage of phosphine from holds, migration, and its dangerous penetration into the residential superstructure of the vessel. Further causes of this problem, above all, were the lack of methods for the pre-trip monitoring of the reliability of cargo hold sealing and, most importantly, the inability to use existing routine methods of finding and eliminating water flow and gas permeability of holds in the state of loaded and phosphine-treated grain cargoes before going to the sea. According to SOLAS, IMO MSC. 79/23 / (2014), in shipping, testing of sealing of empty holds is carried out with the help of water, light, chalk, smoke, ultrasound (Vervloesem,W, 2013:).

However, existing samples which cannot be applied in the practice of transportation of fumigated goods. The effectiveness of practical solutions, using the new test-detector-phosphine method developed by the authors, was supported by merchant fleet experts in 87 - 96% of cases (Belobrov, E, 2016-2017).

Analysis of the causes of poisoning of sailors - 84 respondents (95.3% of cases) phosphine revealed absence for the period of transportation of particular hazardous fumigated cargo, special kits to provide medical aid in cases of poisoning by phosphine, as a required supplement to the regular pharmacy of the vessel for the duration of the voyage. Navy specialists and cargo carriers unequivocally appreciated the importance of the problem and positively assessed the recommendations on the mandatory pre-trip supply of regular shipboard medical aid equipment (SOLAS-74) and provision of the crew with a first-aid kit for phosphine poisoning on the board (Belobrov, E, 2016).

4 CONCLUSION AND RECOMMENDATION

The results of investigations on the causes of accidents and cases of phosphine fumigation gas fatal poisoning of sailors engaged in transporting grain and feed cargo in the ship's holds in an environment of extremely dangerous and toxic fumigants 4.3, 6.1 hazard classes according to IMO IMDG in a voyage allow us to make some conclusion and recommendation:

1 In case of accidental leakage of phosphine from the holds causing accidents, poisoning and death of crew members during sea transportation of fumigated goods, most incidents occur in the group of old, not technically prepared "River-Sea" vessels, where the total amount of pesticides used is not very high.

2 Despite the warning of navigators about the presence in the holds of the ship a cargo that is transported in the environment of poisonous gas (phosphine), in the incident of poison of the crew during the "gas attack fumigation gas", the navigation officers, in many cases are not able to correctly identify the diagnoses. And in the effort to carry out the medical assistance, confusing the symptoms with food poisoning, acute respiratory illness, and poisoning by an unknown gas.

3 The main cause of accidents and acute poisoning of seamen during the transportation of fumigated goods is the lack of training and appropriate qualifications of navigators, especially coastal fumigators, freely passed with toxic chemicals on board the vessel and not having evidence of their competence in the field of knowledge of safety requirements for marine fumigation of cargo.

4 For the prevention of accidental leakage of phosphine and contamination superstructure by

poison, a compulsory practice has been proposed to be carried out by the fumigation team: a monitor of the reliability of the sealing of holds, to detect phosphine leaks by means of a "phosphine detector" invented by us (Ukrainian Patent UA 116604 U) and detection of toxic substances in the vessel's superstructure with the preparation of the relevant Act.

5 Recommended methods of sealing, condensate drain from drainage pipes of the coaming of the hold during the transportation of fumigated goods on ships in a voyage (Patent of Ukraine UA 128846 U).

6 The "Special Medical Chest at Phosphine Poisoning on Board Of the Ship" (Patent of Ukraine for the invention of UA 111028 C2) designed and proposed for the use of the sea fumigators and seaman. In order to improve the efficiency of a rendering of first AID before medical assistance in case of phosphine poisoning on board.

REFERENCES

Belobrov, E., Repetey, V., 2006. The accident on the m/v «Odisk» - non-technological leakage of toxic gas phosphine from the holds and the death of sailors during transportation of ferrosilicon. *Bulletin of the State Fleet Inspectorate of Ukraine, 2006, № (43)*, 117-120.

Belobrov, E., Rangaev,A.Kurbanov,V. 2016. Technology of mandatory control of holds sealing and phosphine effluxes. *Abstr. of the 10th Inter. Conf. on Controlled Atmosphere and Fumigation in Stored Products,(CAF -2016), New Delhi, India, 6-11 November 2016, CAF Permanent Committee Secretariat, Winnipeg, Canada*, 110-111.

Belobrov, E., 2007.*Medical and environmental –hygienic problems of life safety during transshipment in port and transportation of dangerous and fumigated goods on ships under operation conditions and emergency situation* (PhD Thesis), Inter. Academy of Ecology and Life Protection Sciences, Sankt-Petersburg, Russia, 2007, 86 p.

Belobrov, E., Kurbanov,V. Rangaev,A., 2017. Phosphine as a test-detector monitoring the reliability of the sealing of cargo holds before the flight. *"Sea Review" The Inter. Maritime Journal of the Nautical Institute of Ukraine, № 4 (68) 2017, 7-8.*

Belobrov, E. 2016. Efficacy of use of international special chest and its national at poisoning with phosphine at board ship. *Journal «Bulletin of Marine Medicine», № 1 (70), 2016*, 15-19.

Belobrov, E., Sidorenko S. 2015. Hygienic features of the working conditions of workers in marine and agricultural fumigation units. *Journal «Bulletin of Marine Medicine», № 4 (69), 2015*, 101-111.

Belobrov, E., Torskiy, V., Oleshko, A. 2014. Instruction for the provision of medical assistance using a special first-aid kit for phosphine poisoning on board a ship. Odessa-Kiev, "Passage", 2014, 16.

ILO- International Labour Organization. 2014. *Guideless for implementing the occupational* safety *and health provisions of the Maritime Labour Convention , 2006.*

IMO International Maritime Organization. 2008. *Adoption of the code of the international standards and recommended practices for a safety investigation into a marine incident (casualty investigation code).* Resolution MSC.255 (84). (adopted on May 2008).

IMO International Maritime Organization. 2008.. *Recommendation on the safe use of pesticides in ships application to the fumigation of cargo holds.* MSC. 1/ Circ. 1264. 27.05. 2008.

IMO International Maritime Organization. 2014. *Maintenance of Bulk Carrier Hatch Covers Standards for Owners Inspection.* MSC. 79/23/ Annex 2 27.11.2014.

Golikov, V., Repetey V. 2013. Emergency marine accident with the crew poisoning of the m/v «Roksolana-1». *National Maritime Search and Rescue System. Odessa, ONMA*, 83-84.

SMAIC – 2015. *Poisoning of the ship crew after the fumigation of cargo in the port Abidjan on 25 and 26 September 2015.* Final report 47/15 m/v "Nefrit" (Poland).

Torskiy, V., Belobrov, E.,Nezavityn, S. 2012. Practical recommendation for the safe transportation of dangerous and fumigated goods in ice navigation conditions *(Edited by prof. E. Belobrov)* Odessa,8-16.

Verveloesem, W. 2013. Risk & quality throughout the Maritime Logistic Chain. *The Inter. Journal of the Nautical Institute "Seaway", November 2013*, 13-14.

Application of the BPMN Models in Maritime Transport

M. Dramski
Maritime University of Szczecin, Szczecin, Poland

ABSTRACT: Business Process Modeling Notation (BPMN) is a very useful tool in process modeling. The universal character of this approach lets also to apply this methodology into other areas of science, businesses etc. In this paper the use in maritime transport is described. First the event log of the ship's example route was created according to the requirements of process modeling (XES standard). Next the BPMN model was created. Finally the conclusions were made and tips for further research were given.

1 INTRODUCTION

Business Process Modeling Notation (BPMN) has become one of the most widely used languages to model business processes. (Drejewicz, 2017) It's supported by many tools vendors (e.g. Aris) and has been standardized.

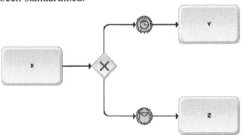

Figure 1. An example of BPMN diagram, source: Drejewicz, 2017

Figure 1 illustrates a very simple process consisting of three tasks. Task marked with the letter Y is launched after the timer sends the signal. Task Z is launched if the message is sent. Besides, in this diagram the XOR-split gateway is present. Naturally, the presented illustration is only the fragment of the BPMN notation. The most common used elements of this language are:
- start event,
- end event,
- intermediate events,
- task/activity,
- AND-split gateway,
- AND-join gateway,
- OR-split gateway,
- OR-join gateway,
- XOR-split gateway,
- XOR-join gateway.

BPMN is supported by Object Management Group – a consortium formed in 1989 which included such companies as IBM, Apple and Sun Microsystems. One of the aims of this consortium is to establish standards for object-oriented programming.

Despite the fact that BPMN is a language intended for modeling business processes, it can be also used in other fields of science and business. It should be noted that business process is nothing but a sequence of certain activities. This leads to the conclusion that this kind of process can be treated in the same way as any other processes. Process modeling aims to understand the real process, but it also have other benefits. Sometimes it allows to increase the efficiency of the process through it's improvement by eliminating the errors, reduction of the execution time or detecting weak points like bottlenecks (the place where the activities have to wait a long time to be executed due to the low process capacity).

On the basis of the above statements, it can be concluded that BPMN models would find it's application also in transport processes. Such example is described in this paper.

2 DATA

Process discovering is the way to find a model of the real process taking place in an organization, company etc. There are several techniques to solve this problem. BPMN models are only one of them. Other examples are: Petri nets, causal nets, process trees etc. Besides each method can be converted to other depending on the current researcher's expectations.

First thing needed in process modeling is of course log data. The data should be given as an event log containing information such: time, activity's name, attributes, resources etc. The most important is always the time. It's natural because the process takes place in a given time period.

To model a process, the data is needed. Usually it should be given in XES data format which is the official standard data structure for process modeling and supporting tools such ProM.

	Case ID	Event ID	dd-MM-yyyy:HH.mm	Activity	Resource	Costs
1						
2	1	35654423	30-12-2010:11.02	register request	Pete	50
3	1	35654424	31-12-2010:10.06	examine thoroughly	Sue	400
4	1	35654425	05-01-2011:15.12	check ticket	Mike	100
5	1	35654426	06-01-2011:11.18	decide	Sara	200
6	1	35654427	07-01-2011:14.24	reject request	Pete	200
7	2	35654483	30-12-2010:11.32	register request	Mike	50
8	2	35654485	30-12-2010:12.12	check ticket	Mike	100
9	2	35654487	30-12-2010:14.16	examine casually	Sean	400
10	2	35654488	05-01-2011:11.22	decide	Sara	200
11	2	35654489	08-01-2011:12.05	pay compensation	Ellen	200
12	3	35654521	30-12-2010:14.32	register request	Pete	50
13	3	35654522	30-12-2010:15.06	examine casually	Mike	400
14	3	35654524	30-12-2010:16.34	check ticket	Ellen	100
15	3	35654525	06-01-2011:09.18	decide	Sara	200
16	3	35654526	06-01-2011:12.18	reinitiate request	Sara	200
17	3	35654527	06-01-2011:13.06	examine thoroughly	Sean	400
18	3	35654530	08-01-2011:11.43	check ticket	Pete	100
19	3	35654531	09-01-2011:09.55	decide	Sara	200
20	3	35654533	15-01-2011:10.45	pay compensation	Ellen	200

Figure 2 An example event log, source: Aalst, 2011

2.1 Preparing the new file with the correct template

Copy the template file B2ProcA4.dot (if you print on A4 size paper) or B2ProcLe.dot (for Letter size paper) to the template directory. This directory can be found by selecting the Tools menu, Options and then by tabbing the File Locations. When the Word programme has been started open the File menu and choose New. Now select the template B2ProcA4.dot or B2ProcLe.dot (see above). Start by renaming the document by clicking Save As in the menu Files. Name your file as follows: First three letters of the file name should be the first three letters of the last name of the first ae.g. e
Open your old file and the new file. Switch between

Figure 2 illustrates a simple event log needed for process modeling. This is of course a fragment of the bigger data set. The most important columns are Case ID, Event ID, timestamp and activity name. At this moment other columns may be omitted but can be used in other analysis.

In this paper the original data recorded during the yacht's voyage from Szczecin to Las Palmas de Gran Canaria is used. The events and the corresponding shortcuts are given in Table 1.

Table 1. The set of events based on the data obtained from Marinetraffic.com, source: Dramski, 2016.

No.	Event name	Shortcut
1	In Range	a
2	Changed course	b
3	Arrival	c
4	Departure	d
5	Stopped	e
6	Underway	f
7	Midnight position	g
8	Midday position	h

The complete event log consists of 59 cases taken during 59 days of the voyage from Szczecin to Las Palmas. The first 30 cases in the form of equation can be given as follows:

$$L = [\langle a,f,c,h,d,e,g \rangle, \langle a,f,b,b,b,b,c,d,g \rangle,$$
$$\langle g,c,e,d,c,h,d,f,c,c,d,a,g \rangle, \langle c,d,a,h,d,a,g \rangle,$$
$$\langle a,h \rangle, \langle g,e,f,b,c,h,g \rangle, \langle h,d,g \rangle, \langle a,h,c \rangle, \langle a,d,h \rangle,$$
$$\langle g,h,c,d,e,f,b,e,g \rangle, \langle a,f,e,g \rangle,$$
$$\langle a,f,e,f,e,c,f,d,h \rangle, \langle g,h \rangle, \langle g,h \rangle, \langle a,h,c,g \rangle, \langle h \rangle,$$
$$\langle g,h \rangle, \langle g,d,h \rangle, \langle g,e,f,c,h,g \rangle,$$
$$\langle h,g \rangle, \langle d,c,e,f,h \rangle, \langle g,c,h,d \rangle, \langle g,e,f,b,b,h \rangle,$$
$$\langle g,e,f,e,f,b,e,f,e,c,h \rangle, \langle g,h \rangle,$$
$$\langle a,f,d,b \rangle, \langle g,h \rangle, \langle g \rangle, \langle g,a \rangle] \quad (1)$$

Now, when the event log is given, the model of the process can be built. In (Dramski, 2016) a Petri net using the α-algorithm is described and in (Dramski, 2017) the same Petri net was created using inductive mining.

3 THE WAYS TO CREATE BPMN MODEL

Creating BPMN models is easy due to the simplicity of this approach (Drejewicz, 2017). Anyway, tools supporting process modeling usually don't have the plugin able to extract the BPMN network directly from the event log.

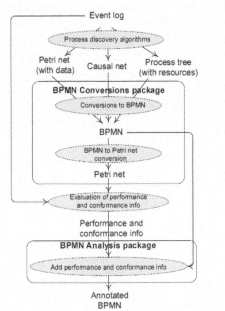

Figure 3 The ways to extract BPMN model from event log, source: Aalst, 2011

Figure 3 illustrates the way to obtain a BPMN model from the event log using one of the most popular tools such ProM, the software developed at Technical University of Eindhoven (The Netherlands).

It is observed that there are three different ways to obtain a BPMN model. The data present in the event log first is used to create Petri net, causal net or process tree. Then the conversion to BPMN is carried out. Next we can convert back this model to Petri net to make possible the evaluation of the model or go straight to BPMN analysis package which allows to extract performance and conformance informations.

Table 2. BPMN plugins in ProM, source: Aalst, 2011

Package name	Plugin name	Funcionality
BPMN Conversions	Convert Petri net	Converts a given Petri net to a BPMN model
BPMN Conversions	Convert Data Petri net	Converts a data Petri net to a BPMN model with data perspective
BPMN Conversions	Convert causal net	Converts causal net to a BPMN model
BPMN Conversions	Convert process tree	Converts a process tree along with resource nodes to a BPMN model with a resource perspective
BPMN Conversions	Convert BPMN model to Petri net	Converts a BPMN model to a corresponding Petri net
BPMN Analysis	Analyze BPMN model	Enhances a BPMN model using performance and conformance information

The example plugins for BPMN models are listed in the Table 2. Anyway, it is necessary to add that ProM software is still developed and new versions are published. Also the plugins are evaluated to eliminate errors and to improve their effectiveness.

4 BPMN MODEL

The data from section 2 of this paper was applied to create a BPMN model. The simple analysis of the XES file is presented in Table 3.

Table 3. The event log statistics, source: own study

Number of processes	1
Number of cases	59
Number of events	216
Number of classes of the events	8
Minimum number of events per case	1
Mean number of events per case	4
Maximum number of events per case	13
Minimum number of event classes per case	1
Mean number of event classes per case	3
Maximum number of event classes per case	7

According to the Table 2 first the Petri net was created.

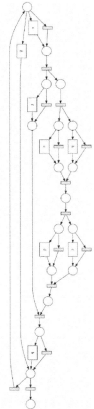

Figure 4. Petri net using inductive mining, source: Dramski, 2017

The Petri net illustrated on Figure 4 was created using inductive mining (Dramski, 2017). Now having the Petri net model it is possible to convert it to BPMN model. So, ProM was launched again and Petri net model was chosen as a resource. Figure 5 illustrates the obtained BPMN model. The conclusion is that this conversion is quite easy. The only significant change is a new notation which is in accordance with BPMN diagrams.

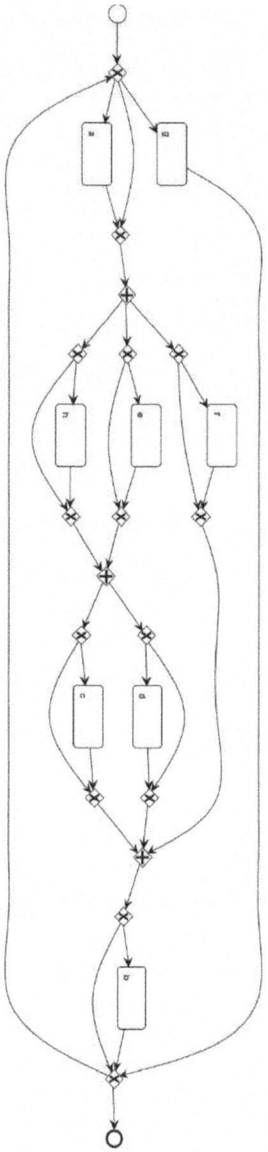

"Simplify BPMN model" and lets to make these model more visible. It is presented in Figure 6.

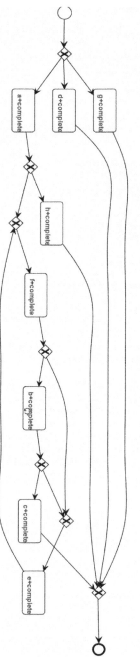

Figure 6. Simplified BPMN model, own study

Figure 5. BPMN model, source: own study

The BPMN model presented above can be simplified using other plugin from ProM. It is called

5 CONCLUSIONS

Comparing the BPMN model and simplified BPMN model it can be observed that the "or" operator was totally reduced. If the Petri net created using inductive mining is taken into the consideration, it has to be mentioned that BPMN model avoids silent transitions, so it also makes this model more clear. Anyway the final conclusion is that this approach to process modelling can be also applied for transport problems. In this paper an example of BPMN model was created using the data presenting the activities during the yacht's voyage from Szczecin to Las Palmas de Gran Canaria (the same data set as in Dramski, 2016 and Dramski, 2017).

BPMN models generally are used for modelling business processes but also can be applied in engineering, IT etc. These models have to be interpreted just as a new notation to create visual representation of the process.

Summing up, the content of this paper indicates that BPMN modelling can be used in transport to improve processes occurring in this area of world economy and the further research in this direction is justified.

REFERENCES

van der Aalst, W.M.P., "Process mining – discovery, conformance and enhancement of business processes", Springer-Verlag Berlin Heidelberg 2011

Dramski, M., "The alpha algorithm in the modeling of the ship's route", Transport Systems Telematics Vol. 9, Iss. 2, p.p. 8-11, 2016

Dramski M., "Inductive mining in modeling of the ship's route", Proceedings of the 12th International Conference, TransNav 2017, 2017

Drejewicz, S., "Zrozumieć BPMN. Modelowanie procesów biznesowych", Wydawnictwo Helion, 2017

Mathematical Model of Hydrodynamic Characteristics on the Ship's Hull for Any Drift Angles

O.F. Kryvyi & M.V. Miyusov
National University "Odessa Maritime Academy", Odessa, Ukraine

ABSTRACT: An effective nonlinear mathematical model of longitudinal and transverse hydrodynamic forces and moment on the hull of the ship for any angles of drift in the case of plane motion is proposed. The construction of the model is based on the physical and mathematical properties of the hydrodynamic forces and uses the hydrodynamic constants of the polynomial models. The model is based on the restoration of the hydrodynamic characteristics of the ship from their Maclaurin series expansions, contains infinitely differentiable functions and simple expressions for the hydrodynamic constants. A numerical analysis of the obtained dependences for the longitudinal, lateral hydrodynamic forces and the moment on the hull of the vessel for different values of the angular velocity and any angles of drift is carried out. Good matching has been established between the proposed model and polynomial models, known nonlinear trigonometric models and experimental data. The proposed model can be written for any type of vessel under any traffic conditions if the hydrodynamic constants of the corresponding polynomial model are known.

1 INTRODUCTION

The study of ship dynamics, in particular, the study of manoeuvres such as braking, acceleration, circulation, Kempf's zigzag, dynamic positioning, etc., presupposes the existence of effective mathematical models of vessel motion that take into account the impact of all inertial and non-inertial forces in wide ranges of motion parameters. Among these forces, the hydrodynamic forces on the hull play an important role.

The study of hydrodynamic forces and the moment on the hull of the vessel devoted a lot of publications. For small drift angles, mostly polynomial models are used. The coefficients of such models are obtained, as a rule, on the basis of the processing of numerical data of full-scale and model tests in wind tunnels, towing tanks, on rotative installations and on planar mechanisms. A detailed description and results of such studies given by Voytkunskiy, Pershytz & Titov (1960); Voytkunskiy (Ed.) (1985); Pershytz (1983); Vasiliev (1989), in particular, coefficients for linear and polynomial models up to the second degree. At the same time, by Fedyaevsky & Sobolev, 1964; Sobolev (1976), using a generalization of the Munk method, a theoretical estimate of the hydrodynamic derivatives of these models was given. Japanese Manoeuvring Modeling Group (MMG) developed by Inoe, Hirano, Kijima, 1981; Yasukawa &

Yoshimura Y, 2015; Furukawa, Ibaragi, Nakiri & Kijima, 2016 are also devoted to building polynomial models. By the method MMG obtained expressions for hydrodynamic derivatives up to the fourth order for different types of ships (Kijima, Katsuno, Nakiri, Furukawa, 1990; Yoshimura, Masumoto, 2012). For large drift angles, there are significantly less adequate models. It should be noted nonlinear trigonometric models (Pershytz, 1983; Voytkunskiy (Ed.), 1985), the hydrodynamic constants of which were obtained for some types of vessels based on the processing of experimental data of model tests. Using the MMG method and the generalization of the Munk method (Fedyaevsky & Sobolev, 1964) by Yoshimura, Nakao, Ishibashi (2009), models of hydrodynamic characteristics were also obtained for a wide range of changes in the drift angle. Sutulo & Soares (2011; 2017) made some interesting generalizations of mathematical models of hydrodynamic forces, in particular, using trigonometric functions, they also gave a detailed description and comparison of well-known empirical mathematical models of hydrodynamic forces and moments.

Analysis of existing models shows that for large drift angles, the calculation of hydrodynamic constants causes considerable difficulties. Expressions for them have a cumbersome appearance and are known for a limited number of types of ships.

The purpose of this work is to build and analyze general adequate simple mathematical models of hydrodynamic characteristics for any drift angles that would be consistent with the physical and mathematical properties of hydrodynamic forces and moment on the hull, and use hydrodynamic constant polynomial models

2 NOTIONS AND GENERAL PROPERTIES OF HYDRODYNAMIC

Among non-inertial forces and moments on the hull of the vessel, special attention is paid to their hydrodynamic components, which are considered when describing any vessel maneuvers. As a rule, the projections X_h, Y_h of these forces on the coordinate axes associated with the vessel, and the moment M_h around the Z axis are expressed as follows

$$X_h = C_h^x(\beta,\omega)\,v^2, Y_h = C_h^y(\beta,\omega)\,v^2, M_h = C_h^m(\beta,\omega)\,v^2. \quad (1)$$

Here v, β, ω – respectively, normalized: the magnitude of the resulting velocity, the drift angle and the angular velocity of the vessel. The values $C_h^x(\beta,\omega)$, $C_h^y(\beta,\omega)$, $C_h^m(\beta,\omega)$ are called hydrodynamic characteristics of the hull. The solvability of the corresponding systems of differential equations of vessel motion causes sufficient smoothness of their right-hand sides, which suggests that the Maclaurin series exist for the hydrodynamic characteristics of the vessel

$$C_h^p(\beta,\omega) = \sum_{j+k=0}^{\infty} C_{jk}^p \beta^j \omega^k, \; \{p\} = \{x,y,m\},$$

$$C_{jk}^p = \frac{1}{(j+k)!} \frac{\partial^{j+k} C_h^p(\beta,\omega)}{\partial \beta^j \partial \omega^k}\bigg|_{\beta=0,\omega=0}, \quad (2)$$

C_{jk}^p are called hydrodynamic constants. Notions (1) allow approximating the hydrodynamic characteristics of ship's hull by polynomials. If, for example, we restrict in the decomposition (1) with terms of infinitesimal order not higher than the third, take into account the consequent equation $C_{jk}^x = 0$, where $(j,k) \neq (0,0)$, $(j,k) \neq (2,0)$, $(j,k) \neq (1,1)$, $(j,k) \neq (0,2)$, $(j,k) \neq (0,4)$, and the equation $C_{jk}^y = 0$, $C_{jk}^m = 0$, where $(j,k) \neq (1,0)$, $(j,k) \neq (3,0)$, $(j,k) \neq (0,1)$, $(j,k) \neq (1,2)$, $(j,k) \neq (2,1)$, $(j,k) \neq (0,3)$, then we receive the following decompositions

$$C_h^x = C_{00}^x + C_{20}^x \beta^2 + C_{11}^x \beta\omega + C_{02}^x \omega^2 + C_{40}^x \beta^4,$$

$$C_h^y = C_{10}^y \beta + C_{01}^y \omega + C_{30}^y \beta^3 + C_{21}^y \beta^2 \omega + C_{12}^y \beta\omega^2 + C_{03}^y \omega^3, \quad (3)$$

$$C_h^m = C_{10}^m \beta + C_{01}^m \omega + C_{30}^m \beta^3 + C_{21}^m \beta^2 \omega + C_{12}^m \beta\omega^2 + C_{03}^m \omega^3.$$

Hydrodynamic constants in the decompositions (6) are expressed through ship's geometric and physical characteristics by processing of experimental data. In this regard, significant amount

of results, considering different external factors and design features of ship's construction such as shallow water, slanting keel etc. Hereunder, for example, frequently used expressions can be emphasized (Kijima et al, 1990; Yoshimura & Masumoto, 2012)

$$C_{20}^x = 1,15 C_B k_2 - 0,18; \quad k_1 = \frac{2T}{L}; \; C_{12}^m = 0,075 k_2^{-1} - 0,098;$$

$$C_{11}^x = 0,08 - 1,91 C_B k_2 + k_1(k_2^{-1} - 0,5); C_{10}^m = k_1; \; C_{00}^x = -C_{x_0};$$

$$C_{40}^x = -6,68 C_B k_2 + 1,1; \; C_{30}^m = -0,69 C_B + 0,66;$$

$$C_{10}^y = 0,5\pi k_1 + 1,4 k_2 C_B; \; C_{21}^m = 1,55 C_B k_2 - 0,76; \; k_2 = \frac{B}{L};$$

$$C_{01}^y = 0,5 k_2 C_B + 0,25 k_1; \; C_{30}^y = 0,185 k_2^{-1} + 0,48;$$

$$C_{21}^y = 0,75 C_B \frac{k_1}{k} - 0,65; \; C_{03}^m = 0,25 C_B k_2 - 0,056;$$

$$C_{12}^y = \frac{0,26}{k_2}(1-C_B) + 0,11; \; C_{02}^x = 0,008 - 0,085 C_B k_2;$$

$$C_{03}^y = 0,058 C_B \frac{k_2}{k_1} - 0,008 \frac{1}{k_1}; \; C_{01}^m = -0,54 k_1 + k_1^2;$$

where T – amidships draft of a ship, C_b – block coefficient, B – waterline beam of a ship, $C_{x_0} = 2R_{x_0}(\rho V_0^2 S)^{-1}$ – coefficient of water resistance to ship's linear motion, $S = LT$, ρ – mass density of seawater ($\rho = 1025 \frac{kg}{M^3}$).

3 EXPERIMENTAL NON-LINEAR MODELS OF HYDRODYNAMIC CHARACTERISTICS

Polynomial models, possessing sufficient simplicity, cover a narrow range of changes in dynamic characteristics of a ship, particularly, for the drift angle the following condition should be fulfilled: $-15° \leq \beta \leq 15°$. Non-linear models containing trigonometric functions correct this deficiency. It should be noted that there is not a big amount of models based on systematic experimental investigations, the most common for positional components are given by (Pershytz, 1983; Voytkunskiy (Ed.), 1985), ($0 \leq \beta \leq \pi$):

$$C_{hp}^x = -\frac{3}{40}\sin((\pi - \arcsin(\frac{40}{3}C_{x_0}))(1-2\sigma_D^4\frac{\beta}{\pi}))$$

$$C_{hp}^y = \frac{1}{2}C_y^\beta \sin 2\beta \cos\beta + c_2 \sin^2\beta + c_3 \sin^4 2\beta \quad (4)$$

$$C_{hp}^m = m_1 \sin 2\beta + m_2 \sin\beta + m_3 \sin^3 2\beta + m_4 \sin^4 2\beta$$

where $\sigma_D = S_D/(LT)$, S_D – reduced area of the submerged part of ships diametrical plane, model parameters $C_y^\beta, c_4, c_2, m_2, m_3, m_4$ are defined according to the nomograms or approximate formulas. As stated in the works (Pershytz, 1983) for pitching (longitudinal) the hydrodynamic characteristics C_{hp}^x, where $0 \leq \beta \leq \pi/2$, the most adequate notion is the following

$$C_{hp}^x = -C_{x_0}\cos\frac{3\beta}{2} - 0.07\sin^4\frac{3\beta}{2} + 8c_4\frac{\beta^3}{\pi^3}. \tag{5}$$

The notions (4), (5) do not cover the whole range of changes in drift angle, particularly for negative drift angles: $\beta < 0$. In the work Vasiliev (1989) there was an attempt to expand the range of its application with the help of non-differentiable functions. The structure of mentioned notions also raises questions. Moreover, the methodology for determining the hydrodynamic parameters of models is quite difficult and inconvenient in application and is achieved for the narrow range of ship's classes.

4 MODELS OF THE HYDRODYNAMIC CHARACTERISTICS FOR ANY DRIFT ANGLES

To eliminate these shortcomings in the representations (4) and (5), it is proposed to apply for constructing models of hydrodynamic forces and moment the approach based on the restoration of the functions (2) (actually (3)) by the expansions that they approximate. Such problems belong to the so-called inverse problems, and in the general case, they are incorrect, because their solutions are ambiguous. Therefore, when applying this approach, it is necessary, first, to consider the physical and mathematical properties of the hydrodynamic characteristics. In addition, it is necessary to perform a numerical comparison of the obtained dependences with experimental data and known mathematical models obtained both for small angles and for large drift angles

For high drift angles, accurate within the third infinitesimal order term, the following correlations are equitable $\beta - \beta^3/6 \sim \sin\beta$, $1 - 0.5 \cdot \beta^2 \simeq \cos\beta$, which allows establishing with the same accuracy, for example, the following correlations $1 \simeq \cos\beta + 0.125\sin^2 2\beta$, $\beta \simeq 0.5(\sin 2\beta + 1/6\sin^3 2\beta)$, $\beta \simeq \sin\beta + 1/6\sin^3\beta$. The last and above-mentioned, properties allow to present the hydrodynamic characteristics of a ship with the decompositions (6) as follows:

$$C_h^x = C_{00}^x\cos\beta + \frac{1}{4}(\frac{1}{2}C_{00}^x + C_{20}^x\cos\beta)\sin^2\beta +$$
$$+ C_{11}^x\omega(\sin\beta + \frac{1}{6}\sin^3\beta) + C_{02}^x\omega^2, \tag{6}$$

$$C_h^y = (C_{10}^y + C_{12}^y\omega^2)\sin\beta + C_{12}^y\omega\sin^2\beta +$$
$$+ \frac{1}{6}(C_{10}^y + 6C_{30}^y + C_{12}^y\omega^2)\sin^3\beta + C_{01}^y\omega + C_{03}^y\omega^3, \tag{7}$$

$$C_h^m = \frac{1}{2}(C_{10}^m + C_{12}^m\omega^2)\sin 2\beta + \frac{1}{4}C_{21}^m\omega\sin^2 2\beta +$$
$$+ \frac{1}{4}(\frac{1}{3}C_{10}^m + \frac{1}{2}C_{30}^m + \frac{1}{3}C_{12}^m\omega^2)\sin^3 2\beta + C_{01}^m\omega + C_{03}^m\omega^3. \tag{8}$$

To check the obtained dependences (6) - (8), their numerical comparison with the experimental data (see Fig. 23 in Sobolev (1976)) and existing models for vessels, whose geometric and physical parameters are given in Table 1, are carried out.

Table 1. Technical characteristics of the ship

	$L(m)$	$B(m)$	$T(m)$	C_b	C_{x_0}	σ_D
Example 1	250	40,8	16,96	0,83	0,018	0,99
Example 2	161	23,2	7,5	0,59	0,01	0,98
Example 3	100	17	4,6	0,64	0,007	0,97
Example 4	320	58	20,8	0,81	0,023	0,99

Fig. 1 and 2 shows a comparison between the hydrodynamic characteristics achieved using the obtained notions (6) - (8) (solid lines) and a polynomial model (3) (dashed lines). Curves 1-3 in Fig. 1 correspond to the Examples 1-3. The curves 1 in Fig. 2 show the dependence $C_h^y(\beta,0)$ and the curves 2 show the dependence $10 \cdot C_h^m(\beta,0)$ for the Example 2. The results show that for all three hydrodynamic characteristics both models provide an acceptable match in the following range: $-15° \leq \beta \leq 15°$, that, in fact, is coherent with the application area of polynomial models.

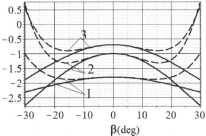

Figrue 1. $100 \cdot C_h^x(\beta,0)$ for the polynomial and obtained models

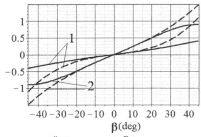

Figure 2. $C_h^y(\beta,0)$ and $10 \cdot C_h^m(\beta,0)$ for the polynomial and obtained models

Fig. 3 and 4 show the dependence of the hydrodynamic characteristics on the drift angle for the obtained notions (6) - (8) (solid line), notions (5) (dashed lines) and the improved notions (6) (dotted line) for Example 2. In particular, Fig. 3 shows graphs for $10 \cdot C_h^x(\beta,0)$ in the range: $0° \leq \beta \leq 90°$, Fig. 4 shows graphs for $C_h^y(\beta,0)$ (curves 1) and (curves 2) in the range $0° \le \beta \le 180°$. The results show the practical match of property values $C_h^m(\beta,0)$ obtained from the formulas (5) and (9). The good consistency of behavior is visible for other characteristics too. At the same time the behavior pattern of the hydrodynamic characteristics $C_h^x(\beta,0)$ and $C_h^y(\beta,0)$ available from experiments (see Fig. 23 in Sobolev G.V. (1976)) can be more accurately described by the dependences (6) - (8). It is especially noticeable for $C_h^x(\beta,0)$ at the angles of $40° \le \beta \le 80°$ and for $C_h^y(\beta,0)$ at the angles of $40° \le \beta \le 140°$.

The advantage of notions (6) - (8) is a wide spectrum of changes in hydrodynamic parameters, particularly, they cover the whole range of drift angle variation. Proposed dependencies contain infinitely differentiable functions. Moreover, hydrodynamic constants of the model (6) - (8) are selected from polynomial model (5), recently the methodology of their determination has been significantly developed.

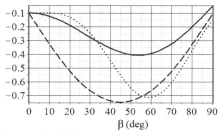

Figure 3. $10 \cdot C_h^x(\beta,0)$ for the obtained and other non-linear models

Thus, the obtained representations (6)-(8) are in good agreement with the existing polynomial and trigonometric experimental models. In addition, the advantage of the representations (6)-(8) is a wide range of changes in the dynamic parameters, in particular, they cover the whole range of the drift angle change. The proposed dependences also contain infinitely differentiable functions, and the hydrodynamic constants of the model (6)-(8) are chosen from polynomial models, the technique of obtaining them has recently developed significantly.

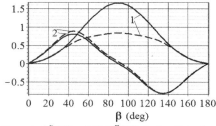

Figure 4. $C_h^y(\beta,0)$ and $10 \cdot C_h^m(\beta,0)$ for the obtained and other non-linear models

5 NUMERICAL ANALYSIS OF THE OBTAINED MODELS

Let us illustrate the nature of the change in the hydrodynamic forces for different values of the drift angle and angular velocity. Fig. 5 and 6 shows the dependence of the hydrodynamic characteristics on the drift angle $\beta \in [-180°, 180°]$ where $\omega = 0$ for the suggested model (6) - (8). The solid lines in both figures correspond to the Example 1, dotted lines correspond to the Example 2, dashed lines correspond to the Example 3 and dot-and-dash lines correspond to the Example 4. Lines 1 in Figure 6 correspond to the $C_h^y(\beta,0)$ characteristic and lines 2 correspond to the $10 \cdot C_h^m(\beta,0)$ characteristic. The obtained results are absolutely coherent with the experimental data.

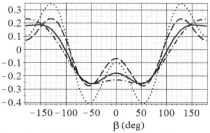

Figure 5. $10 \cdot C_h^x(\beta,0)$ for the obtained model

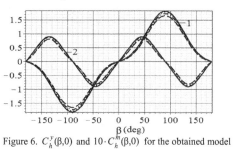

Figure 6. $C_h^y(\beta,0)$ and $10 \cdot C_h^m(\beta,0)$ for the obtained model

Fig. 7-12 for the Example 1 show the dependence of the hydrodynamic characteristics of the obtained model (7) - (8) for the fixed values of the non-dimensional angular velocity ω (Fig. 7, 9, 11) and for fixed values of the drift angle β (Fig. 8, 10, 12). In particular, Fig. 7 shows the dependence of the hydrodynamic characteristic $C_h^x(\beta,\omega_j)$ on the drift angle $\beta \in [-180°;180°]$ for the fixed values of the non-dimensional angular velocity: $\{\omega_j\} = \{\pm 0.5, \pm 1, \pm 1.5, \pm 2\}$ Fig. 9 shows the dependence of the hydrodynamic characteristics $C_h^y(\beta,\omega_j)$, $C_h^y(-\beta,-\omega_j)$ for the values $\{\omega_j\} = \{0.1, 0.3, 0.6, 0.9\}$. Fig. 11 shows the dependence of the hydrodynamic characteristics $C_h^m(\beta,\omega_j)$, $C_h^m(-\beta,-\omega_j)$ for the fixed values: $\{\omega_j\} = \{0, 1, 1.5, 2\}$. Fig. 8 shows the dependence of the hydrodynamic characteristic $C_h^x(\beta_j,\omega)$ on the velocity ratio $\omega = [-3;3]$ for the fixed values of the drift angle: $\{\beta_j\} = \{0,\pm30°,\pm45°,\pm60°,\pm90°\}$. Fig. 10 shows the dependence of the hydrodynamic characteristic $C_h^y(\beta_j,\omega)$ for the fixed values $\{\beta_j\} = \{0,\pm30°,\pm45°, \pm60°,\pm90°\}$ Fig. 12 shows the dependence of the hydrodynamic characteristics $C_h^m(\beta_j,\omega)$, $C_h^m(-\beta_j,-\omega)$ for the values $\{\beta_j\} = \{0,20°,60°,80°\}$. The obtained results are coherent with the results of works obtained with the use of polynomial models at slight drift angles $(-15°\le\beta\le15°)$ and illustrate the adequacy of the suggested model.

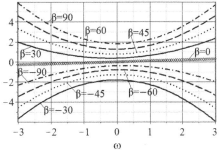

Figure 10. $C_h^y(\beta_j,\omega)$ for the obtained model

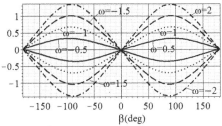

Figure 7. $C_h^x(\beta,\omega_j)$ for the obtained model

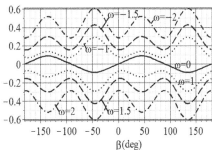

Figure 11. for th $C_h^m(\beta,\omega_j)$, $C_h^m(-\beta,-\omega_j)$ e obtained model

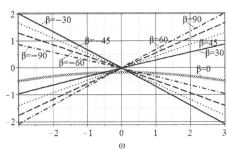

Figure 8. $C_h^x(\beta_j,\omega)$ for the obtained model

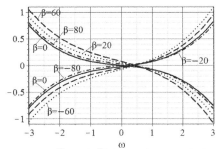

Figure 12. $C_h^m(\beta_j,\omega)$, $C_h^m(-\beta_j,-\omega)$ for the obtained model

Three-dimensional diagrams of dependencies $C_h^x(\beta,\omega)$, $C_h^y(\beta,\omega)$, $C_h^m(\beta,\omega)$ are described in Fig. 13-15 with the aim of visual illustration of the hydrodynamic characteristics (6) - (8).

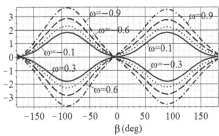

Figure 9. $C_h^y(\beta,\omega_j)$, $C_h^y(-\beta,-\omega_j)$ for the obtained model

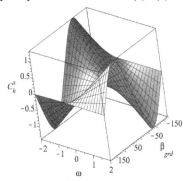

Figure 13. $C_h^x(\beta,\omega)$ for the obtained model

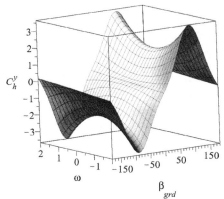

Figure 14. $C_h^y(\beta,\omega)$ for the obtained model

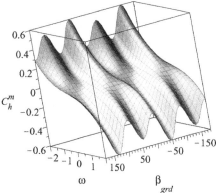

Figure 15. $C_h^m(\beta,\omega)$ for the obtained model

6 CONCLUDING REMARKS

Thus, an effective model of the hydrodynamic characteristics of the ship with simple expressions for the hydrodynamic constants was obtained. This mathematical model covers the entire variation range of the drift angle and is fully coherent with the existing polynomial and non-linear models, as well as with the experimental data. The suggested model can be used for any type of ship under all operating conditions if the hydrodynamic constants of the corresponding polynomial model are known. It should be noted that the proposed approach to the construction of mathematical models of hydrodynamic characteristics allows further development due to the introduction of additional parameters and improvement of approximation expansions in the model.

REFERENCES

Fedyaevsky KK, Sobolev GV. 1964. Control and Stability in Ship Design. Washington, DC: US.

Furukawa Y, Ibaragi H, Nakiri Y & Kijima K. 2016. Shallow water effects on longitudinal components of hydrodynamic derivatives. Uliczka et al. (Eds). Proceedings of 4th International Conference on Ship Manoeuvring in Shallow and Confined Water with Special Focus on Ship Bottom Interaction (4th MASHCON) pp. 295-303. Karlsruhe: Bundesanstalt für Wasserbau. DOI: 10.18451/978-3-939230-38-0_33.

Inoe S, Hirano M, Kijima K. 1981. Hydrodynamic derivatieves on ship maneuvering. Int. Shipbuilding Progress 28(321):112-125. DOI: 10.3233/ISP-1981-2832103.

Kryvyi OF. 2015. Methods of mathematical modeling in navigation (in Ukrainian). Odessa: ONMA.

Kryvyi OF, Miyusov MV. 2016. Mathematical model of movement of the vessel with auxiliary wind-propulsors. Shipping & Navigation 26: 110-119 (in Russian) http://nbuv.gov.ua/UJRN/sudovozhdenie_2016_26_18.

Kijima K, Katsuno T, Nakiri Y, Furukawa Y. 1990. On the manoeuvring performance of a ship with the parameter of loading condition. J Soc Nav Archit Jpn 168: 141–148. https://doi.org/10.2534/jjasnaoe1968.1990.168_141.

Miyusov MV. 1996. Modes of operation and automation of motor vessel propulsion unit with wind propulsors (in Russian). Odessa: OSMA.

Pershytz RY. 1983. Dynamic control and handling of the ship (in Russian). Leningrad: Sudostroenie .

Sobolev GV. 1976. Dynamic control of ship and automation of navigation (in Russian). Leningrad: Sudostroenie .

Sutulo S & Soares Guedes C. 2011. Mathematical Models for Simulation of Maneuvering Performance of Ships. In Guedes Soares et al. (Eds.). Marine Technology and Engineering (pp 661–698) London: Taylor & Francis Group. DOI: 10.13140/2.1.3538.7209

Sutulo S and Soares Guedes C. 2017. Comparative simulation of definitive manoeuvres of the KVLCC2 benchmark ship using different empiric mathematical models. Guedes Soares & Teixeira (Eds.). Maritime Transportation and Harvesting of Sea Resources (pp.572-579). London: Taylor & Francis Group. ISBN 978-0-8153-7993-5

Vasiliev AV. 1989. Dynamic control of vessels (in Russian). Leningrad: Sudostroenie.

Voytkunskiy Ya I, Pershytz RY & Titov IA. 1960. Handbook on Ship Hydrodynamics: Resistance, Propulsion, and Manoeuvring (in Russian). Leningrad: Sudostroenie.

Voytkunskiy YaI (Ed.) 1985. Handbook of the ship theory. In 3 volumes (in Russian). Leningrad: Sudostroenie.

Yasukawa H, Yoshimura Y. 2015. Introduction of MMG standard method for ship maneuvering predictions. J Mar Sci Technol 20:37–52. DOI: 10.1007/s00773-014-0293-y.

Yoshimura Y, Nakao I, Ishibashi A. 2009. Unified Mathematical Model for Ocean and Harbour Manoeuvring. Proceedings of International Conference on Marine Simulation and Ship Manoeuvrability (MARSIM2009), pp.116-124 http://hdl.handle.net/2115/42969.

Yoshimura Y, Masumoto Y. 2012. Hydrodynamic Database and Manoeuvring Prediction Method with Medium High-Speed Merchant Ships and Fishing. Proceedings of International Conference on Marine Simulation and Ship Manoeuvrability (MARSIM 2012), Singapore, 23 April, pp.494-504.

New Passenger Maritime Transport System for Gulf of Cadiz (ESPOmar PROJECT)

J.I. Alcaide, R. García Llave, F. Piniella & A. Querol
University of Cádiz, Cádiz, Spain

ABSTRACT: ESPOmar Project, co-financed by INTERREG POCTEP 2014-2020 program of the European Regional Development Fund (ERDF) together with the Spanish Universities of Cadiz and Huelva, Portuguese University of Algarve and Andalusian Agency of Public Ports, is a cooperation research network aimed at the theoretical design of a new sustainable cross-border maritime transport system for the Gulf of Cadiz, helping to conserve and protect the environment, promote sustainable development and protect the natural and cultural heritage of the fluvial coastal and maritime area, especially the National Park of Doñana which has been threatened on the last decades by the installation of high capacity highways. The main challenge of ESPOMAR project is to analyze the real demand of such a transport system, and optimize the design of specific ferries that may compete with the stablished land transport system in terms of speed, safety and comfort, reducing emissions and assuring the economic viability of potential maritime lines. On the other hand, ESPOmar project is also trying to stablish the interest that this kind of transport system could have for tourist population, not only in terms of improving their mobility, but also for increasing their leisure offer with an attractive new way of connecting tourist areas in such a way that passengers are able to admire the wild fauna, the fantastic beautiful coast and natural environment of the Gulf of Cadiz. The aim of this paper presented is to introduce ESPOmar Project in an international forum, to show the first results and conclusions of our research study and receive comments and inputs from the scientific community.

1 INTRODUCTION

The object of the research project to be developed jointly and coordinated by the different participating research centers, is focused on the littoral zone of the Gulf of Cádiz and the Guadiana River. The very physiognomy of the Gulf, as well as the existence of the Doñana Natural Park (JA 2009), make the road connections much longer than the potential maritime connections, so that the maritime distance between Cadiz and Faro is 156 kilometers shorter than the distance by road. Also, the distance between Cadiz and Huelva by sea would be 105 kilometers shorter than by road (Figure 1).

Doñana National Park is one of the most important wetlands of Europe and emblem of biodiversity and the protected areas of the Spanish region of Andalusia (Granados 1987).

The mouth of the Guadalquivir River is an estuary that includes part of the main channel of the river and about 14 miles of coastal strip in the coast of Doñana. It is characterized for being a dynamic system, rich in nutrients and high productivity, where its importance lies in: being an area of laying, breeding and fattening of fish, molluscs and crustaceans of great fishing interest, influence the fisheries of the Gulf of Cádiz, constitute a migratory channel for many species, and contribute to sustaining part of the Doñana ecosystem (BOJA 2004).

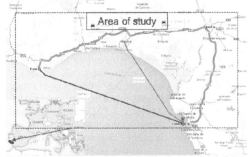

Figure 1. Location map of the Gulf of Cádiz (Area of study).

For decades, sectors of Cadiz and Huelva, two bordering provinces with no direct road contact and separated by the Guadalquivir River and the National Park, have been claiming for a direct r. To go from one capital to another you are forced to travel North and pass o through Seville city and (with traffic jams in the round that means). Although the first voices spoke of a road along the coast, some political initiatives even bet on the north variant: unfold a small current road that borders the natural park (greater than the national) by the north and then build a new and complex bridge over the Guadalquivir. This new road could have an environmental impact from a point of view of protection of the natural space, in addition to public expenditure; isolating the protected area completely and turning Doñana into a kind of zoo rather than a natural area. On the other hand, this measure, even from an economic point of view, would be against other policies such as the promotion of tourism in national parks.

In the mid-seventies begins to speak of the construction of a coastal road between Huelva and Cádiz that, among other difficulties, had the ability to cross the Guadalquivir at its mouth. The Government did not include the construction of this highway in its infrastructure plan. The existence of other motorways in service or execution and environmental requirements discourage their construction, but that does not mean that both provinces continue to call for an improvement in connections and that are even political options that put the construction of these highways before environmental criteria.

Today we can find innumerable systems of maritime transport that regularly connect cities that are surrounded by rivers, bays, etc., in addition to the maritime lines that connect the different islands to each other or to the continent (Piniella et al., 2014). In the area of study, sea passenger transport has provided service in a traditional way, although not continued in different areas of the Gulf of Cadiz, and on both sides of the border between Spain and Portugal, as an alternative to private vehicles, both the Bay of Cádiz as the western coast of Huelva. The line of maritime transport of travelers between Ayamonte and Vilareal de Santo Antonio (Portugal) annually exceeds one hundred thousand passengers and more than two thousand vehicles along the Guadiana River. There is a maritime line Huelva-Punta Umbría, with more than ten thousand travelers. It has a passenger pier in the port of Punta Umbria (remodeled, with a public investment of € 420,000) by the Ministry of Development and Housing. Another line joins the port of El Rompido with the Flecha beaches, crossing the Ría del Piedras, as well as a boat-bus service that connects the port of Isla Cristina with that of Isla Canela (Ayamonte). Without forgetting, in the province of Cádiz, the metropolitan maritime service of the Bay, which covers by catamaran the route between Cádiz and El Puerto de Santa María and between Cádiz and Rota.

The Society demands that transport systems be more respectful with environment, more efficient and socially sustainable. The proposed project will address different aspects aimed at facilitating mobility under criteria of efficiency, operability, economic and environmental sustainability between coastal and riparian populations located in the Spanish provinces of Huelva and Cádiz and in the Portuguese Algarve. Likewise, it is intended to design a transport system that supposes a tourist attraction through which to take advantage of the natural and cultural potential of the area, thus contributing to a greater and better economic development.

1.1 *Hypothesis*

The development of ESPOmar project, from the point of view of the sustainable transport system, will hopefully help to protect, promote and develop the cultural and natural heritage of the Gulf of Cadiz, as well as to reduce atmospheric pollution. As a working hypothesis, 3 possible navigation routes are being considered to connect the three bordering regions of Algave, Huelva and Cadiz. Different alternatives of ports, direct and multimodal transport for each connection are being also analyzed. In successive phases of expansion of the project, it would be necessary to study the possible connections within Guadiana river.

1.2 *Aim and scope*

The aim of this paper presented is to introduce ESPOmar Project in an international forum, to show the first results and conclusions of our research study and receive comments and inputs from the scientific community.

2 THE ESPOMAR PROJECT

The ESPOmar project (Spain and Portugal joined by the sea) is led by the University of Cádiz, with a multidisciplinary approach where the Universities of Huelva and the Algarve are integrated, together with the Andalusian Ports Agency. The research proposed by the project under the name "Network of Cooperation in R + D + i Orientation to the Design of a Sustainable and Transboundary System of Maritime-Fluvial Transport in the Gulf of Cádiz" is co-financed by INTERREG POCTEP 2014-2020 program of the European Regional Development Fund (ERDF) of the European Union. The proposed project was granted an ERDF budget allocation of € 272,286.86. The term of execution of the contract began on October 2017 and will end on December 31, 2019.

The University of Cádiz considers that ESPOmar project as the best way to strengthen the Campus of International Excellence of the Sea (CEIMAR), having as its objective the creation of a cooperation network in R & D between different centres. The research involves to the Universities of Algarve in Portugal and those of Huelva and Cadiz in Spain. Likewise, the object of study of the ESPOmar Project, maritime and fluvial transport in the Gulf of Cádiz, is considered as a public subject necessary for a sustainable development of the area, improving the tourist offer, the use of the patrimonial resources, together with a possible reduction of greenhouse gas emissions.

This communication is an initial presentation of our project, where we will try to address some of our preliminary proposals; however, these cannot be considered definitive until the completion of the different phases that make up the project.

The underlying objective of the ESPOmar Project is based on the establishment of a Cooperation Network in R + D + i Orientation to the Design of a Sustainable and Transboundary System of Maritime-Fluvial Transport in the Gulf of Cádiz, which helps to improve cross-border connectivity, the management of natural resources, the conservation, protection, promotion and development of the natural and cultural heritage, contributing to the increase of the tourist offer, and in the economic and sustainable growth of the zone of fluvial coastal maritime activity included between Cádiz and Faro. A series of specific activities were established for this study:
– Study of demand;
– Maritime lines definitions;
– Design of sustainable Boats;
– Environment effect investigation;
– Economic analysis and optimization.

Of the objectives proposed in the project, they hope to obtain the following results:
– Creation of a cross-border R & D & I cooperation network to allow coordinated solutions to common problems in the area;
– Design of a sustainable system of maritime and fluvial transport;
– Innovation in the design of transport vessels, optimized for the uses of each transport line and in the most sustainable way, with less polluting energy sources and with less impact on the natural environmental;
– Analysis of the demand of local and tourist population;
– Feasibility and economic impact study;
– Environment effect investigation;
– Analysis and optimization of results.

In an initial phase, a study of the demand has been made in the main population centers that surround the Gulf of Cádiz. To later integrate the demand obtained with the design of the possible maritime lines considered in the study area. In the section of the maritime lines it contemplates the regulations applicable to the maritime transport of passengers in the planned lines; a navigation study has been prepared on aspects related to the bathymetry, effects of current and wind, speeds, beaconing, interferences on the routes with the density of maritime traffic or fishing activity in the area.

Next, the design of the prototype of the vessel will be analyzed, considering that society demands transport systems to be increasingly respectful of the natural environment, efficient and socially sustainable. As a working hypothesis, possible navigation lines that surround the Gulf of Cádiz are presented. Highlighting the lines proposed by the project, the great landscape value, cultural and gastronomic resources of the populations that are linked to them.

2.1 Project coordination

The execution of a project of such magnitude requires an organization structure with a well-defined distribution of functions and activities, in this sense, the Monitoring Committee is coordinated with the person responsible for redirecting the general orientation of the project (implementation schedule and its adjustments, plan of activities, joint tasks and actions, presentation of payment certifications and reports, monitoring and evaluation of the execution).

In the operational section, the Thematic Commissions will be in charge of the management and monitoring of project activities, as well as their transferability, focus on the results of the thematic tasks, focus on the efforts of the partners in the execution of Your actions, can be organized by the following subgroups:
1 Commission for Preparation, Management and Coordination: Send the bases of the studies to be carried out, ensure its correct execution and work for the sharing of information and documents related to its execution;
2 Commission for the Study and Design of the Transport System;
3 Commission for the Design of Vessels and Port Infrastructures;
4 Commission for Study the Environmental Impact;
5 Economic Analysis and Optimization Commission;
6 Information and Publicity Commission: It will define the communication actions of ESPOmar as well as the Coordination with the activities of the partners that are complementary.

Regarding the work methodology, it is intended to facilitate, on the one hand, the effective and efficient management of the project and, on the other hand, to reinforce the cross-border dimension. It is

based on the exchange of information and experiences, on the form, data, solutions, common problems and specific problems. Its basic principles are:
- Participation, both in decision-making and in the orientation of the project;
- Decentralization, to favor the freedom of maneuver in the execution of the entrusted tasks, within the framework of the common limits;
- Flexibility, to adopt decisions that respond to common interests;
- Adaptation to the type of activity.

3 STAGES OF THE PROJECT

From the perspective of the activities that take place during the execution of the project, a series of stages have been planned, which imply a system of execution as a whole. The system or methodology is intended to be the structure that feeds back the results of the activities and those of the project as a whole (Figure 2). In the following sections will be treated the different phases of the project, which are integrated by: Definition of the demand; Design of routes; Integration with other modes of transport; Environmental viability of the routes; Optimization of the boats; Analysis of economic viability; Economic impact in the area; Optimization of the maritime transport system.

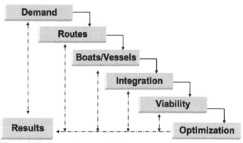

Figure 2. Design a transport system.

The work methodology will allow us to develop a guide of the various activities of the project, on the other hand, they provide a frame of reference in the follow-up of the results and an optimization of the resources. The different information blocks and workflows that organize the design and validation of maritime routes, in addition to the adaptation of the vessels to the conditions of service.

3.1 *Shipping routes*

The maritime transport network discussed in the project would be integrated by 3 maritime lines, where the longest distance to navigate does not exceed 160 km. In order to determine the best ports,

passenger capacity, number and service speed of ships, 10 maritime lines and 8 ports are being analyzed and compared in terms of both environmental and economic impact. Determinants for a successful route operation:
- Good layout (peculiarity of each route or area, surveys, stakeholders, etc.);
- Government assistance (initial period);
- Vessels used (sustainable, comfortable, fast, etc.);
- Port infrastructure & interfaces (Standardized, efficiency, facilities, etc.);
- Promotion (routes, service, area, culture, etc.).

3.2 *The vessel design*

In the geographic space of reference proposed by the project, the main handicap posed by the design of the vessels is direct competition with already stablished land transport vehicles (train, coaches and cars). The starting point is determined by transit times between populations. Figure 3 shows the workflows that determine the sustainable design of the vessels.

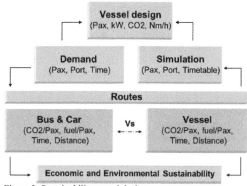

Figure 3. Sustainability vessel design.

One of the main challenges presented by the project is the adaptation of the vessels to the characteristics of the demand, that is, to the number of passengers per route along the route and to the total service times (including boarding and disembarkation of the passage). On the other hand, the conditions of the own navigation in open waters must be kept in mind, among others: sea state, maximum wave height, balance periods, vibrations, noises, etc. The main axis of the design would be defined by the environmental impact of the boats and a second axis of reference would be formed by the particular characteristics that define the comfort and safety of the passengers.

3.3 *The economic study*

The economic block of the project contemplates a detailed study of the economic feasibility of exploitation of the maritime lines proposed by the same, along with the construction costs of the vessels, personnel cost, operating costs, industrial benefit, and the determination of the price of the ticket (Stopford 2013).

In the economic impact section, the direct and indirect jobs created by the activity will be treated in depth, among others; the economic impact derived from communications; the economic impact derived from the tourist attraction; and the social benefits of transport (de Rus et al., 2006).

4 CONCLUSIONS

This communication is not only a report of our research project, but also a presentation to the scientific community. We hope that it is well received and accepted by the Regional Public Administrations, as well as the provincial and municipal authorities of Cádiz, Huelva and the Algarve, which will be responsible for adopting and promoting the future passenger service by sea in case this research stablishes that is environmental and economic sustainable

Likewise, it is intended to design a transport system that supposes a tourist attraction through which to take advantage of the natural and cultural potential of the area, thus contributing to a greater and better economic development.

The project would create an integrated passenger transport network. This would provide various benefits to the population, such as reducing traffic congestion, reducing costs and saving time for travelers on these routes, along with benefits for the natural environment.

The transport system that is intended to design may reduce traffic substantially in the area of Seville, therefore the environmental impact study will determine the final figures of the transport system not only from the maritime point of view but for the whole integrated system included the actual land transport system.

Espomar Project research is not promoting maritime connections instead of land transport system. We are trying to determine if it is possible to design a new maritime transport system that could help to the environmental and economic sustainability of the Gulf of Cadiz. However, we will not know if this is possible until the conclusions of the research project are stablished.

DISCLAIMER

The authors would like to thank the European Regional Development Fund (ERDF) of the European Union (EU) for co-financing the project "ESPOmar" under the "INTERREG POCTEP 2014-2020 program". We also like to thank to all researchers and staff of the different public organizations involved on the project for their dedication and professionalism.

BIBLIOGRAPHY

BOJA, 2004. Boletín Oficial de la Junta de Andalucía (BOJA), ORDEN de 16 de junio de 2004, por la que se declara una Reserva de Pesca en la desembocadura del río Guadalquivir. Boletín número 123 de 24/06/2004. Junta de Andalucía.

De Rus, G., Betancur, O., Campos, J. (2006). *Manual de evaluación económica de proyectos de transporte.* Banco Interamericano de Desarrollo. Washington, D.C.

Granados Corona, M. (1987). *Transformaciones históricas de los Ecosistemas del Parque Nacional de Doñana.* (Tesis doctoral inédita). Universidad de Sevilla, Sevilla.

JA, 2009. Junta de Andalucía (JA). *Mapa Guía Digital de Espacios Naturales: Doñana Espacio Natural.* Consejería de Medio Ambiente. Seville: JA.

Piniella, F., Querol, A. & Rasero, J.C. (2014). Maritime passenger transport as an urban and interurban alternative on the river Guadalquivir: GuadaMAR In Martinez, X (ed.), *Maritime Transport VI – 6th International Conference on Maritime Transport:* Barcelona. 443-452.

Querol, A., Jiménez-Castañeda, R., Piniella, F. (2013). Department Ship Design Optimization Applied for Urban Regular Transport on Guadalquivir River (GuadaMAR). [in:] A. Weintrit & T. Neumann (eds.) *Information, Communication and Environment – Marine Navigation and Safety of Sea Transportation.* London, UK. 179-184.

Stopford, M. (2013). *Maritime economics.* 3ed. Routledge.

Ship Design Challenges for ESPOMAR Project: A Review of Available Methods

M.J. Legaz, A. Querol & B. Flethes
University of Cádiz, Cádiz, Spain

ABSTRACT: ESPOMAR is a European Interreg Project (POCTEP 2014-2020) in which several universities, research groups and public organizations participate. The basic idea of ESPOMAR is to study the possibility of establishing a competitive regular maritime passenger line between Spain and Portugal in the Gulf of Cadiz. The project includes also the design of the ships optimized for each of the defined maritime lines. Having in mind that the new maritime transport system we pretend to stablish must compete with the existing land transport based on private car and public buses, several challenges arise in our ship design process that must be solved. In one hand, fast ships need to be used to compete with land transport time, but in the other, we need to assure that the ship are efficient and eco-friendly, using a sustainable propulsion systems and hydrodynamic optimized hulls. Likewise, we need also to deep analyze the ship dynamic behavior so that passengers feel as comfortable as they do in the land transport. State of art in ship dynamic damping systems should also be analyzed to define the best option for our ferries. A review of the methods and bibliographic available is made in this paper to find out the best option to meet the challenges of ESPOMAR project.

1 INTRODUCTION

ESPOMAR is the acronyms of the project name "España and Portugal unidos por el mar (Spain and Portugal connected by sea)". This project has been obtained in the framework of European Interreg Projects (POCTEP 2014-2020). The aim of ESPOMAR project is to study the possibility of stablishing a competitive regular maritime passenger line between Spain and Portugal within the Gulf of Cadiz.

The Gulf of Cadiz is located in the southwest part of Iberian Peninsula and is surrounded by the Spanish provinces of Cadiz and Huelva and the Portuguese region of Algarve. In the figure 1 and 2, the interest area and maritime connections can be seen. The characteristics of the enclave, the existence of the Guadalquivir River and Doñana National Park, make the road connections much longer than the potential maritime connections. In fact, the maritime distance between Cadiz and Faro is 156 kilometers shorter than the distance by road and the necessary to travel north and pass throw out Seville allow us thinking that waterborne transport could be a more sustainable transport option. The main idea is to establish lines between Cadiz, Huelva and Algarve, with the possibility to establish others fluvial lines within river Guadiana, which is a natural border between the two countries.

Partners of ESPOMAR Project are the Spanish Universities of Cadiz and Huelva, Portuguese University of Algarve and Andalusian Agency of Public Ports. One of the scope of the project is the design of the ships optimized for each of the defined lines. The new ship designs have to be able to compete with the existing land transport based on private car and public buses, to accomplish the regulations and to follow the European Union transportation trends.

Figure 1. Area of interest. Source: Google Maps

Figure 2. Maritime connections. Source: Google Earth

Ship design has to face with several challenges to accomplish the previous one. These challenges are: the ships have to compete in time with land transport, they need to be efficient and eco-friendly, the passengers have to feel comfortable.

A review of the methods available to meet these challenges will be made in the next sections. In the section 2 a revision of the type of fast ships for passenger transport is made. In the section 3, some eco-friendly European projects and eco-friendly catamarans are shown. Some remarks about the hull optimization process is made in section 4. The ways of obtain more comfortable catamarans is explained in section 5.

2 TO COMPETE WITH LAND TRANSPORT

In order to compete with the land transports, the ships have to be fast. The ships that operate between islands or in coastal lines use to be classified in the category of "fast ferries", among them can be found: hydrofoils, hovercrafts, trimarans, catamarans and monohulls. Nowadays, the trend in fast ferries designs are catamarans and monohulls.

The authors have carried out some statistical study of ferries with various capacity, service speeds and range of operations, in order to analyze the relation between the installed propulsion power (BHP) and number of passenger and speed requirements. Figure 3 shows the installed propulsion power (kW) per passenger ratio versus service speed (dark line is the exponential tendency line).

In short-sea shipping passenger ferries, catamarans are quite popular between operator because of their high speed, low construction and maintenance cost. Around 60% of high speed ferry market belong to catamarans [18]. Regarding to speed, several catamarans are in service with speeds of 35-40 knots and displacements from a few hundred tons to around 3,850 tons. In sheltered waters, small catamarans have reached speeds superior to 50 knots [11].

Others advantages of catamaran design are:

– Low wave-making because of high length/beam ratio. The designer can minimize the propulsion engine power rating for a given service speed.
– High transverse stability due to the space between hulls. The transverse metacentric height (GM) will be around ten times higher than a monohulls.
– Large deck area, because of the centre bridge between the side hulls, providing spacious and comfortable passenger cabins and other working cabins. Deck area in catamaran is greater about 40-50% than monohulls.
– High manoeuvrability and course stability due to the space between twin propulsors giving a larger turning moment.
– Low impact and slamming load s as well as speed loss in waves due to hull slenderness compared with monohulls.
– Subdivision against flooding. The catamaran provides a high safety level against hull damage because of many bulkheads in both side hulls with small individual compartment volume. This has as result smaller flooding in case of damage [18].

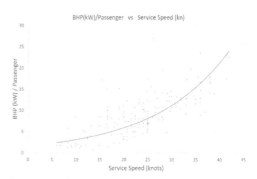

Figure 3. BHP (kW) per passenger ratio versus service speed.

3 ECO-FRIENDLY PROPULSION SYSTEMS

The global warming due to greenhouse effect has received the attention of authorities during the last years. The increment of the warm in the earth will have many implications on ecosystems and in the life of human being.

The government of different countries have made an effort for reduction of CO_2 emissions. These emissions affect to the maritime transport. IMO established an index to measure the CO_2 emissions in new ship, the EEDI index. The energy efficiency design index determines how clean is a ship. It can be expressed as the ratio of environmental cost divided by benefit for the society: EEDI= impact to environment/benefit to society. In others words, EEDI= CO_2 emissions/transport work [9]. EEDI index was made mandatory for new ships and the

Ship Energy Efficiency Management Plan (SEEMP) for all ships at MEPC 62 (July 2011) with the adoption of amendments to MARPOL Annex VI (resolution MEPC.203(62)), by Parties to MARPOL Annex VI [12].

This index depends on the type of ship and the size, some values of the EEDI index for passenger catamaran can be found in reference [17]. Eco-friendly propulsion systems, in this paper, means systems of propulsion that reduce emissions of CO_2 and obtain good values of EEDI index.

Several European project and research works have been made in order to obtain eco-friendly ships. The lines of investigation have gone in the sense of obtain: less pollutant marine engines, to use batteries in the short-shipping travels, combine LNG (Liquid Natural Gas) and diesel engines and use renewable energy systems combine with engines. Also some others techniques such as using hydrogen cell can be found. Some of the European projects are the following:

The E-Ferry project has the aim to reduce CO_2 emissions and air pollution from marine transport. The project pretends to demonstrate the feasibility of 100% electrically powered of medium size ferry for passengers, car, trucks and cargo. These ferries usually operate in island communities, coastal zones and inland waterways. The battery-pack or the E-Ferry are one of the largest ever installed in a ferry. They have high charging power capacity of up to 4MW. This allows the stays in the port be short. An image of the batteries is shown in Figure 5. Other aspects of the design and construction are being taking into account in order to make the E-Ferry efficient. Some of them have an optimized hull-shape, lightweight equipment and carbon composite materials [3]. An image of E-Ferry is shown Figure 4.

Figure 4. E-Ferry. Source:[3]

The E-ferry reaches higher speed at restricted depths and in confined waters. The sailing velocity will be 13-15 knots instead of 10-12 knots [3]. Other projects like E-Ferry and SEABUBBLE, BB Green, GFF worked on the development of electric

ferries and small electric boats for passenger transport.

Another European project is the Green Fast Ferry (GFF) is a full-electric, high speed passenger craft. The objective of this project was developed a fast ferry with speed of 30 knots and zero emissions. The ferry has large autonomy, 26 km in 30 minutes and the recharge at port is fast, less than 20 minutes. GFF use the technology of Air Supported Vessel (ASV), based on a pressurized air cushion underneath the hull that reduces energy consumption in at least 40%. ASV are only appropriated for being used in sheltered or inland waters [7].

Figure 5. Batteries of the E-Ferry project. Source:[3]

Some of fastest ferries that are currently sailing and combine engines and LNG (Liquefied Natural Gas) are:

The LNG-powered-high speed catamaran from the Spanish shipping company Balearia. The Balearia ferry has a length of 125 meters and a beam of 28 meters. The capacity will be 1,200 passengers and 500 cars being one of the longest and highest-capacity fast ferry catamaran. The propulsion system will be four Wärtsilä dual LNG/diesel engines delivering 8800 kW each. The catamaran will have a service speed of 35 knots, and a top speed of over 40 knots. Its range will be of 400 nautical miles, being equipped with two tanks to store the chilled fuel. Balearia Company has the aim to extent LNG ferries along their lines in the Mediterranean. The first eco line is established in Baleares islands. The project has received a rating of excellent from de European Union [1]. In the Figure 6 can be seen the catamaran Balearia.

Francisco, a high-speed Ro-Ro ferry built by Incat for its operator Buquebus, operates with liquefied natural gas (LNG) as the primary fuel. Francisco (formerly Lopez Mena) is the first dual fueled high-speed Ro-Ro ferry to operate with liquefied natural gas (LNG) as the primary fuel. The high-speed ferry was built by Incat for its operator Buquebus. It started to operate in 2013, on the River

Plate, between Buenos Aires, Argentina, and Montevideo, Uruguay.

Figure 6. Balearia catamaran. Source: [1]

The catamaran Francisco has a length of 99 m, a beam of 26.94 m. It has capacity for 1000 passengers and 150 cars. The propulsion system consists of two Wartsila LJX 1720 SR waterjets driven by two GE Energy LM2500 gas turbines. A water jet and gas turbine place into each hull. Four Caterpillar C18 generators rated at 340kW each, and two Caterpillar C9 generators rated at 200kW each also can be found as auxiliary components.

The maximum speed of Francisco is of 51.8kt with a deadweight capacity of 450t. The catamaran has been classified as 1A1 HSLC R4 Car Ferry B Gas Fueled E0 [5]. In the Figure 7 an image of catamaran Francisco it is shown.

Figure 7. Francisco catamaran. Source: [5]

Other options for eco-friendly fast ferries are the use of hydrogen fuel cell technologies [6] and the hybrid diesel electric propulsion system [8]. The use of conventional system of propulsion with sources of renewable energy is also another eco-friendly option. Renewable power applications can be used in ships of all sizes for primary, hybrid and/or auxiliary propulsion.

The sources of renewable energy that can be used in ship are wind, solar-photovoltaics, wave energy and biofuels.

Regarding to wind, the energy can be captured using soft-sails, fixed-sails, flettner rotors, kite sails and wind turbines.

- Soft-sails are conventional sails attaching in yards and masters. This is a mature technology that harness directly the propulsive force of wind.

- Fixed-sails can be described as rigid wings with a rotatory mast. Some projects are UT Wind Challenger [16] and Effship's [4].
- Flettner Rotors using the Magnus effect that produce the wind passes over an already revolving cylinder, to obtain propulsion.

- Kite sails placed in the bow of the ship can operate to the appropriated altitude to maximise wind speeds.
- Wind turbines have been installed for ship propulsion over the years. Until now, they are not really successful.

Solar energy use electricity generated by photovoltaic cells. These cells can be placed in different ways onboard; in fixed wings sails or horizontally on deck [10]. There are several options to be eco-friendly, during the development of ESPOMAR project one of them or a combination of them will be selected as the best option for the project.

4 EFFICIENT SHIPS

The optimization process in a ship can be applied from different perspectives. In this section, efficient ships refer to a ship in which its hull forms offer the minimum resistance to shipping through the sea water.

The resistance the ship experiment by sailing through the sea water, depends in an important part of its hull form. The process used to design and defined the hull form is for this reason quite important. At the beginning, the way of verifying the resistance in the ships are tank tests with models. The resistance of the proposed model is measured. The dimensional analysis is used in order to obtain the resistance of the ship. With the advance of the computers and its calculation power, the computational fluid mechanics (CFD) is used to study the characteristics of the fluid around the hull of a ship [13]. Although, tank tests are still necessary to bear out the computers results.

Nowadays, the tendency followed in several research of different countries is the combination of computer aided design (CAD), computational fluid mechanics (CFD) and optimization methods [20]. The figure 8 shows the cyclic process.

Regarding to CAD, there are different ways to introduce the hull in a CAD software depending on the type of CAD software that is been used. The hull can be introduced in CAD software in a parametric or non-parametric way (conventional design). In a non-parametric way, the hull is introduced by a set of points. Depending on the type of software, the software use NURBS, B-Spline, etc. as interpolation technique to adjust the hull form. The software have some fairness criterium. In the parametric way, the hull is defined by a set of variables and parameters. In the parametric design the changes can propagate thought the model updating all related parts, therefore only a few modifications are required in order to achieve a new fair hull form.

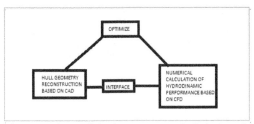

Figure 8. The optimization process of hydrodynamic performance based on CFD technology.

Respect to optimization, different decisions have to be taken. If the problem is a single-object or multi-object. In a single-object only one component of the problem has to be minimized. In the multi-object problem several components of the problem have to be minimize.

When we face of a multi-object problem a process of making decisions has to be adopt since usually an improvement in a component carries out a deteriorate of another component of the problem. The model is not able to combine the best performance in every objective so the decision-making plays a critical role in optimization. The aim of the decision process is to find the best compromised solution. This solution can be found in the pareto front, obtaining a set of feasible outcomes. Also, several numerical algorithms can be used to solve the optimization problem.

As an example of a research work developed in this area is: the work of Zaraphonitis and others [19], in which a hull form optimization procedure is develop in order to minimize wash and total resistance of high speed ships. The work was based on the integration of three software: NAPA a ship design software package was used for the generation of several hull forms. The CFD software "Shipflow" to perform the hydrodynamic evaluation of each hull form. The optimization of the hull was made with mode Frontier software, using the method of Genetic algorithms. This work studied two types of ships a semi-displacement monohull and a high-speed catamaran.

5 COMFORT FOR PASSENGERS

Another important aspect that ESPOMAR project has to reach is comfort for their passengers. Passengers can suffer the vertical and horizontal accelerations provoked by the sailing of the ship,

this cause discomfort and dizziness. In the case of catamarans, the profile rolling and pitching natural periods are very similar. This provokes significant discomfort to the passengers and crew, especially in the case of shipping in quartering seas [18].

One thing that can be done is modified the hull form in order to reduce vertical acceleration, combined with roll and pitch motions. In this sense the work of Piscopo and Scamardella [14] can be seen. In the work, the Motion Sickness Incidence (MSI) is applied to the hull form optimization of a wave piercing high-speed catamaran vessel. The hull is modelled in a parametric way to generate two families of derived hull forms, the former varying the prismatic coefficient and the position of longitudinal center of buoyancy. When the optimum hull is generated, vertical accelerations at some critical points on main deck are com-pared with the parent ones. The MSI index is defined as the percentage of passengers who vomit after two hours of exposure to a certain motion.

The problem of modified the hull form to obtain comfort is that probably the new hull form is not efficient. It is difficult to change the pitch and roll natural period without generating an inefficient hull form [18]. For this reason, attention has put on changing the damping coefficient of the hulls in roll and pitch.

In order to increase damping, automated pitch control foils and transverse anti-rolling fins have become widely used on modern high-speed catamarans and do improve high-speed catamaran seaworthiness significantly. Other elements used to mitigate the movement in the ship are anti-rolling tanks and bilge keels.

As example can be seen the work of De la Cruz and others, improving the comfort of a fast ferry. This research focused on a ship sailing on head seas. They considered a ship with two flaps at transom and a T-foil near the bow, the control is applied to them in order to improve the comfort [2].

6 CONCLUSIONS

Recalling that it is a preliminary study, the main conclusions obtained are:
− In ships design as in most of the engineering design, a compromise solution must be taken among the different objectives and requirement of the project.
− In this work, a revision of the methods and systems currently available and that can help to meet the challenges of ESPOMAR project have been made.
− The work has focused in catamaran fast ferries, because it is thought to be the most appropriate ship for the project. A revision of European project, hull optimization methods and damping

systems that can be used in this type of ships have been made.
- At the end the responsibility of deciding what are the most appropriate technological development and the methodologies, is of the design team.

DISCLAIMER

The authors would like to thank the European Regional Development Fund (ERDF) of the European Union (EU) for co-financing the project "ESPOmar" under the "INTERREG POCTEP 2014-2020 program". We also like to thank to all researchers and staff of the different public organizations involved on the project for their dedication and professionalism.

REFERENCES

[1] Balearia Catamaran. https://www.lngworldnews.com/spains-balearia-building-lng-powered-high-speed-catamaran/ (online access 08/01/2019).
[2] De la Cruz J.M., Aranda J., Girón-Sierra J.M. & Velasco F.J. 2004. Improving the comfort of a fast ferry. IEEE Control Systems Magazine.
[3] E-ferry Project. https://cordis.europa.eu/project/rcn/193367/factsheet/es (online access 08/01/2019).
[4] Effship's Project. http://www.effship.com/Public Presentations/Final_Seminar_2013-03-21/06_EffShip-Wind_propulsion-Bjorn_Allenstrom_SSPA.pdf (online access 08/01/2019).
[5] Francisco Catamaran. https://www.ship-technology.com/projects/francisco-high-speed-ferry/ (online access 08/01/2019).
[6] Genessis Ferry Project. https://www.ship-technology.com/projects/hydrogenesis-passenger-ferry (online access 08/01/2019).
[7] GFF Project. https://cordis.europa.eu/project/rcn/206526/factsheet/en (online access 08/01/2019).
[8] Hybrid Ferry Project. https://www.ship-technology.com/projects/mv-hallaig-hybrid-ferry (online access 08/01/2019).
[9] Indian register of shipping. Implementing Energy Efficiency Design Index (EEDI). Powai
[10] IRENA. 2015. Renewable energy options for shipping.
[11] Lamb R. Robert Lamb 2003. High-speed, small naval vessel technology development plan. West Bethesda.
[12] MARPOL. https://www.marpol-annex-vi.com/eedi-seemp/ (online access 08/01/2019).
[13] Min K.S., Choi J.E., Yum D.J., Shon J.H., Chung S.H. & Park, D.W. 1999. Study on the CFD application for VLLC Hull-Form design. Twenty-Second Symposium on Naval Hydrodynamics.
[14] Piscopo V., Scamardela A.2015. The overall motion sickness incidence applied to catamarans. International Journal of Naval Architecture and Ocean Engineering.
[15] Tzannatos E. & Stournaras L. 2014. EEDI analysis of Ro-Pax and passenger ships in Greece. Maritime Policy & Management.
[16] UT Wind Challenger Project. http://wind.k.u-tokyo.ac.jp/index_en.html (online access 08/01/2019).
[17] Walsh C., Bows A. 2012. Size matters: Exploring the importance of vessel characteristics to inform estimates of shipping emissions. Applied Energy.
[18] Yun L. & Bliault, A. Springer 2012. High Performance Marine Vessels. New York
[19] Zaraphonitis G., Papanikolaou A., Mourkayannis D. 2014. Hull-form optimization of high speed vessels with respect to wash and powering. Conference paper.
[20] Zhang B.J. & Zhang S.L.2019. Springer. Research on Ship Design and Optimization Based on Simulation-Based Design (SBD) Technique. Shanghai.

Study on Applying Numeric Modeling CFD for Fuel Injection Process of Common Rail System in Marine Diesel Engine

V. Quan Phan & H. Dang Tran
Ho Chi Minh City University of Transports, Ho Chi Minh City, Vietnam

ABSTRACT: Today, the use of common rail technology for high power marine diesel engines is becoming more and more widespread. The outstanding advantage of the electronic injection system is the in-creased efficiency, engine power and NOx reduction in engine emissions, and the use of electronic injection systems will increase reliability and performance of the engine. To study and analyze the outstanding features of the electronic injection system, the author examines the application of CFD numerical simulation to develop a program to simulate the working process of a fuel injection of marine diesel engine. This program is used to analyze the factors affecting the fuel injection process of the marine diesel engine nozzle, which gives the researcher a basis for evaluating the optimization features of the system. Electronic injection is al-so used as a tool for analyzing breakdown failures of electronic fuel injectors during operation. In this presentation, the author will in turn introduce the electronic fuel injection theory, calculation and simulation program, CFD simulation technique in Ansys and run the program with specific applications for analyzing and evaluating the quality of fuel injection process.

1 COMMON RAIL SYSTEM IN MARINE DIESEL ENGINE

1.1 *Common rail system:*

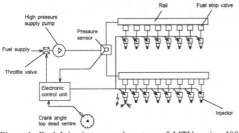

Figure 1. Fuel Injection control system of MTU series 4000 (Johannes Kech et al, 2011)

Common rail system was developed in 1913 by Vickers company in England (P.G.Burgman and F. Deluca, 1962) but until 1995, this system had been applied for diesel engine by DENSO in Hino Rising Ranger Truck (Miyaki, M. et al, 1988). Since then, the common rail system has been popular and used largely.

In a common-rail engine, a feed pump delivers the fuel through a filter unit to the high-pressure pump. The high-pressure pump delivers fuel to the high-pressure accumulator (the rail). The injectors are fed from this rail. The injectors inject fuel into the combustion chamber when the solenoid valve is actuated. The fuel volume between the high-pressure pump and the injectors serves as an accumulator. This helps to dampen oscillations initiated by the pulsating delivery of the high-pressure pump. A pressure sensor measures the fuel pressure in the rail. Its value is compared to the desired value stored in the electronic control module (ECM). If the measured value is different from the desired value, an overflow orifice on the high-pressure side of the pressure regulator is opened or closed. The overflow fuel returns to the tank. The injector opening and closing is controlled by the ECM. The duration of injection, fuel pressure in the rail and the flow area of the injector determine the injected fuel quantity. (Stumpp and Ricco, 1996)

1.2 *Common Rail system in marine diesel engine*

According to the advantages of the common rail system, it has been applied in marine diesel engines broadly so far. Major marine engine manufacturers have participated in this model such as MAN B&W, Sulzer, Caterpillar, Cummins, MTU, etc. This paper limits the research for studying the high-speed diesel engine using common rail system. The system in marine sector is quite different with that used in automotive, due to its operation condition on water with wave and wind.

Figure 2. SAR413 ship equipped common rail MTU engine [1]

In order to have the optimum condition to research, this study focuses on the popular engines equipped the common rail system on marine ships in Vietnam. MTU engines are chosen due to their widespread and standby condition for research. These engines are series 4000, with the L'Orange injector (Johannes Kech et al, 2011), installed in the rescue fleet of Vietnam Maritime search and Rescue Coordination Center.

2 FUEL INJECTION PROCESS

2.1 Parameters of common rail injection

When researching the common rail system, there are several parameters that should be investigated, such as injection pressure, injection timing and the forming and developing of the injection spray.

Pressure: Fuel injection pressure of a common rail system is always maintained at the set value and not affected by the various speeds of the engine. Global automotive supplier DENSO Corporation has developed a new diesel common rail (DCR) fuel injection system with the world's highest injection pressure of 2,500 bar. Based on DENSO's research, the new system can help increase fuel efficiency by up to 3 percent while also reducing particulate matter (PM) by up to 50 percent and nitrogen oxides (NOx) by up to 8 percent. This is compared to DENSO's previous generation system. (Denso Website)

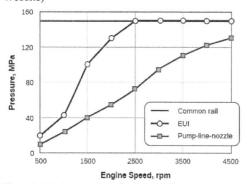

Figure 3. Injection pressure of common rail system compared with others (Denso Website)

Injection timing: the injection timing is controlled by the Electronic Control Module (ECM). This ECM controls the piezoelectric fuel injectors in terms of opening or shutting them off. Injection timing can be varied during running of the engine, whereas in conventional system the engine has to be stopped and setting for timing has to be changed. (Krogerus, T. & Huhtala, K, 2018).

Injection spray: Currently, it is possible to have up to eight injection pulses - two pilot, four main, and two post-injection pulses during the injection of a common rail system (Mohamad R. H. et al, 2016). By using two high-speed video cameras to capture the development of the injection spray, it is stated that by increasing injection pressures and using small injector orifice sizes, the injected diesel droplets reduce in size and penetrate further, hence increasing air utilization, thus leading to faster evaporation rates and reduced ignition delay (Figure 4).

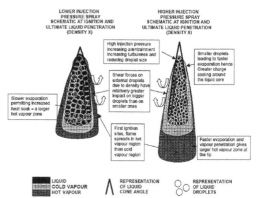

Figure 4. Summary schematics of two spray developments in similar in-cylinder densities for different injection pressures. (Mohamad R. H. et al, 2016)

2.2 Injection process of common rail system

In common rail system, the injection process occurs following 3 phases: pre-injection, main injection and post injection. When the fuel is sprayed into the combustion chamber of the engine, each fuel particle is subject to simultaneous effects of internal force and external force. Internal forces include the bonding force between molecules, external tension. External forces include the initial excitation force when traveling through the injection orifice and aerodynamic drag of compressed air in the combustion chamber. The bonding force between molecules and external tension depends heavily on the molecular structure, viscosity and surface tension, which tends to keep the fuel injection beam continuous, not shredded. Meanwhile the initial dynamic force and the aerodynamic resistance of compressed air in the combustion chamber tend to tear the fuel beam into smaller particles. In addition, when circulating fuel through the injection orifice will occur disturbance due to the relatively high traveling speed, the degree of turbulence of the spray beam depends mainly on the viscosity, surface tension of the course. Under the impact of internal forces as well as external forces, resulting in the disintegration of fuel beams into particles of different sizes, shapes and concentrations. The particles are small because the value of internal forces is much smaller than the impact of external forces.

The process of disintegrating the fuel injection beam is formed immediately after the fuel is sprayed into the engine cylinder. The separation of the fuel injection beam takes place quickly or rapidly depending on factors such as the spray velocity, pressure and temperature of the combustion chamber, the physical properties of the fuel. The decay process is an important prerequisite for breaking down the fuel injection beam into small particles, which include primary decay and secondary decay.

During primary decay, fuel droplets and particles are separated from the spray beam. These drops and particles of fuel are often unstable and therefore can continue to decay. After the primary decay process is a secondary decay process. These two decay processes are critical to the geometric parameters of the fuel injection beam and the size of the fuel particle.(D.A. Taylor, 1996).

2.3 *Impacts of basic parameters of common rail injector to injection process*

In order to evaluate the spray forming and quality, there are basic parameters of the injector should be considered. These are the quantity of the hole, nozzle type (SAC or VCO), and diameter of the orifice. In order to ensure the quality of forming mixed gas in Diesel engine cylinder, it is necessary to ensure fuel spray into small sized particles, each about 5-50 micrometers in diameter.

The smaller size of the fuel particles is, the greater the total heating and evaporation area of a unit of fuel volume is, so the faster the evaporation of fuel particles. In Diesel engine, the time for creating mixture is very short from 0.05 to 0.001 seconds. Therefore, in order to ensure the good combustion of the fuel, it must be evenly distributed in the volume of the combustion chamber, the mixture must be uniform, the fuel rays must ensure sufficient distance, must be able to penetrate the air layer loaded.

The fineness of the fuel particles is expressed through the average diameter of the particles contained in the fuel. The higher the number of revolutions, the shorter the time of creating the mixture requires more fine spraying and smaller

spraying. In order for the process of forming mixed gas with good quality, it is necessary to ensure the speed of the fuel passing through the spray holes from 150 to 400 m / s. The velocity of fuel rays increases when the injection pressure increases. When the injection pressure increases, the average diameter of the particles decreases. The farther away from the nozzle hole the lower the velocity of the beam. Increasing the injection pressure will increase the distance of the beam while increasing the air pressure in the engine cylinder when the constant injection pressure will reduce the length of the fuel beam. Increasing the orifice diameter will increase the beam length but will also increase the average diameter of the fuel particles. (Dough, W., 2004)

2.4 *Conclusion*

Basing on the above parameters, this study chooses to research on the diameter and form of the orifice of the common rail injector. The injector after a long time working under very high pressure will be worn. This can be gradually worn in round shape, or in any shapes. Thus, it causes several effects to the forming of the spray beam after injection.

In order to do this, the authors have applied the Ansys Fluent software to simulate the forming of the diesel fuel spray beam when going out of the common rail injector with different models and circumstances of the orifice. And this will provide the direct view in detail of the quality of the spray.

3 APPLYING CFD FOR FUEL INJECTION PROCESSOS COMMON RAIL SYSTEM

3.1 *Short introduction CFD and applications*

CFD - Computational Fluid Dynamics: This is a field of science using numerical methods combined with computer simulation technology to solve problems related to environmental movement factors, physical characteristics of processes under consideration, environmental strength properties, thermodynamic properties, kinetic properties, or aerodynamic properties, force characteristics, or material force and the interaction of the environment together, ... depending on each object and the specific scope of each problem, each scientific area that the CFD can be applied.

ANSYS FLUENT is a software with the ability to widely model physical properties for fluid flow, turbulence, heat exchange and reaction modeling applied in the industry from flow through the wing fly to a fire in a furnace, from gas bubbles to oil cushions, from the flow of blood vessels to the construction of semiconductor materials and from the design of clean rooms to the processing equipment wastewater treatment. Special models make the software capable of modeling the engine combustion chambers, acoustic aerodynamics, wing machines and multi-phase systems to serve the expansion of software capabilities.

3.2 Mathematic modeling of injection process

The spray after going out of the nozzle of the injection process can be considered as the free flow going through the very small hole. However, the fuel after nozzle will combine with the compressed air inside the chamber, thus the mixture contains 2 phases: fluid and gas. For this problem, there are 3 equations applied:
- Continuity equation of mixture:

$$\frac{\partial}{\partial_t}(\rho_m) + \nabla.(\rho_m \vec{v}_m) = 0 \tag{1}$$

In which $\vec{v}_m$ is the mass averaged velocity

$$\vec{v}_m = \frac{\sum_{k=1}^{n} \alpha_k \rho_k \vec{v}_k}{\rho_m}$$

ρ_m is density of the mixture

$$\rho_m = \sum_{k=1}^{n} \alpha_k \rho_k$$

α_k is volume fraction of phase k
- Momentum Equation:

$$\frac{\partial}{\partial_t}(\rho_m \vec{v}_m) + \nabla.(\rho_m \vec{v}_m \vec{v}_m) = -\nabla\rho + \nabla.\left[\mu_m\left(\nabla\vec{v}_m + \nabla\vec{v}_m^T\right)\right] + \tag{2}$$
$$\rho_m \vec{g} + \vec{F} + \nabla.\left(\sum_{k=1}^{n} \alpha_k \rho_k \vec{v}_{dr,k} \vec{v}_{dr,k}\right)$$

where n is the number of phases, $\vec{F}$ is a body force, μ_m is the viscosity of mixture

$$\mu_m = \sum_{k=1}^{n} \alpha_k \mu_k$$

$\vec{v}_{dr,k}$ is the drift velocity for secondary phase k:

$$\vec{v}_{dr,k} = \vec{v}_k - \vec{v}_m$$

- Volume fraction equation for air phase:

$$\frac{\partial}{\partial_t}(\alpha_p \rho_p) + \nabla.(\alpha_p \rho_p \vec{v}_m) = -\nabla.(\alpha_p \rho_p \vec{v}_m) + \sum_{q=1}^{n}(\dot{m}_{qp} - \dot{m}_{pq}) \tag{3}$$

Reference: ANSYS Fluent 15 Theory Guide - 17.4. Mixture Model Theory – page 500

3.3 Programing CFD in Ansys for common rail injection process

To investigate the outcome of the fuel spray beam going out of the injector of the common rail system, the model of this simulation has following conditions:

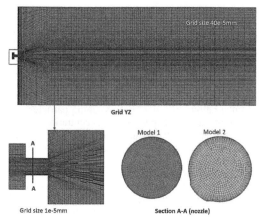

Figure 5. Grid declaration for 2 models (works from authors)

Table 1. Parameters and conditions of the model.

Specifications of the injector	
Diameter of orifice	0.15mm – 0.16mm
Number of orifices	01
Injection Pressure	1,500 bar

Specifications of the mixture		
	Phase 1 Diesel oil	Phase 2 Air
Density (kg/m³)	730	Ideal air
Specific Thermal Capacity (J/kg.K)	2090	1010
Viscosity (kg/m.s)	0.001584	2.138e-5
Thermal conductivity (W/m.K)	0.149	0.0306

Table 2. Specifications of the model grid

	Cells	Faces	Nodes
Model 1	540,850	1,627,473	545,911
Model 2	610,014	1,835,071	615,179

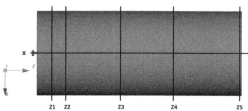

Figure 6. Definitions of the surface of combustion chamber (works from authors)

For the Model 1, the orifice diameter is 0.15mm – as the popular diameter of the injectors of L'Orange brand for MTU engines. In the Model 2, the orifice is assumed to be worn so the diameter is increasing to 0.16mm together with a small defect. The actual diameter of the cylinder is 165mm will be modified to 6mm to reduce the computational operating time length but still ensure that it does not affect the actual conditions of the spray beam.

3.4 Result modelling by CFD

Ansys Fluent is used in this case to demonstrate the forming of the velocity of the injector at the nozzle (Section X), as well as describe the mixture volume fraction at this point. Then, the mixture volume fraction at Section Z (from Z1 to Z5) will be simulated to show the forming of the mixture after the nozzle and into the combustion chamber. The simulation also shows the view of the forming and spreading the fraction volume of the mixture of fuel and air when go pass each section.

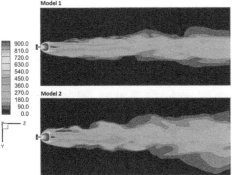

Figure 7. Velocity distribution at Section X (works from authors)

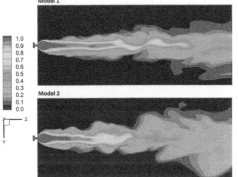

Figure 8. Mixture volume fraction at Section X (works from authors)

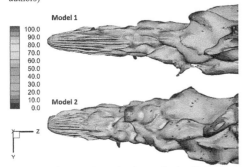

Figure 9. Mixture volume fraction at Section Z (works from authors)

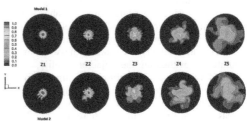

Figure 10. Mixture volume fraction at Section Z1 – Z5 (works from authors)

3.5 Analyzing modeling process

The flow velocity at the output of the nozzle is very high (~ 800 m/s), this extremely high velocity create an air shock (compressed air zone with a high compression ratio) at the output. This is what causes the fuel spray beam to be misty.

3.5.1 Comparison between 2 models:

Model 2 (corresponding to the case of the nozzle being abraded and having a defect corner) shows that the fuel flow is unstable after just getting out of the nozzle, besides creating vortices, while model 1 (corresponding to the case of a normal nozzle, not abrasive), the fuel flow is relatively stable after getting out of the nozzle and the flow is far from a new distance to gradually form the vortex.

In model 2, the more dispersed fuel line at the quadrant of the nozzle is missing, which is the main reason for instability after the fuel escapes the nozzle.

Actually, when the authors try to increase the diameter of the orifice, the forming of the spray does not have the clearly differences, but when there is a defect on the round shape of the orifice, the turbulence occurs visually.

Therefore, it can be proved that if the nozzle is worn at the uniform (still keeping the round shape), the injector can be utilized continuously, and the quality of the fuel spray is adequate. However, if there is any defect happens on the round shape, the injection quality need to be reviewed and the injector should be rectified in order to ensure the quality of the burning process of the engine.

4 CONCLUSION

The common rail fuel system on marine diesel engine is very different with the traditional fuel system, especially about the injection pressure. Under high pressure of the fuel going out of the nozzle, the injector needs to be manufactured for the durable operation. The Ansys Fluent has work very well to show the visually model and to simulate the forming of the mixture between diesel oil and compressed air in order to demonstrate the quality of the mixture inside the combustion chamber before being burnt. However, this is the simulation for only one factor of the injection process – the diameter of the orifice individually. In order to have the more accurate in evaluating the injector quality, there should have been a combination simulation of various sectors. For example, what happens if the orifice has defect and the injection pressure is increased; or if the injector has more than one hole. These cases make the authors have to do more research in the future to figure out the various circumstances of defects and conditions, in order to establish some related functions for the rectifications of the common rail injectors towards different types of defects.

REFERENCES

ANSYS Fluent 15 Theory Guide - 17.4. Mixture Model Theory – page 500

D.A. Taylor, 1996. "Introduction to Marine Engineering" (Second Edition). Elsevier, Oxford.

DENSO Global Website. "DENSO Develops a New Diesel Common Rail System With the World's Highest Injection Pressure - News.

Dough, W, 2004. "Pounder's Marine Diesel Engines and Gas Turbines", 8th edition, Elsevier UK.

Johannes Kech et al, 2011. "Common Rail Fuel Injection: Key technology for clean and economical combustion" – MTU General White paper 2011.

Krogerus, T. & Huhtala, K, 2018. "Diagnostics and Identification of Injection Duration of Common Rail Diesel Injectors". Open Engineering, 8(1), pp. 1-6. Retrieved 30 Jan. 2019, from doi:10.1515/eng-2018-0001

Miyaki, M. et al, 1988. "Fuel Injection System", US Patent 4,777,921

Mohamad R. H. et al, 2016. "Characteristics of pressure wave in common rail fuel injection system of high-speed direct injection diesel engines" – Advances in Mechanical Engineering 2016, Vol. 8(5) 1-8. SAGE

P.G.Burgman and F. Deluca, 1962. "Fuel Injection and Controls: for Internal Combustion Engines", Technical Press, London.

Stumpp and Ricco, 1996. "Common Rail-An attractive fuel injection system for passenger car DI Diesel Engines", SAE Paper 960870.

The International Convention for the Control and Management of Ships' Ballast Water and Sediments: A Respectful Way to Tackle the Problem of Aquatic Biodiversity in Maritime Transport

R. García Llave & J.I. Alcaide
University of Cádiz, Cádiz, Spain

ABSTRACT: In order to mitigate the effects caused by harmful aquatic organisms and pathogens contained in ships' ballast water, the International Convention for the Control and Management of Ships' Ballast Water and Sediments established the obligation to implement on board a ballast water management system as well as operational limits of depth and distance while sailing off the coast.

This piece of work analyses, having as a reference the Strait of Gibraltar, the main provisions of this Convention applicable to the discharge of ballast water, as well as the general practice established in the Regional Agreement for the management of ballast water in the Mediterranean Sea. This paper also analyses the methodology that Spain should follow in case that the country pondered the possibility of designating a maritime zone for the transfer of ballast water in this geographical area.

1 INTRODUCTION

The use of ballast water by the search engines when they sail totally or partially without cargo in the holds is a common practice that allows them to maintain safe operational conditions during the trip, since efforts are reduced on the hull, transverse stability is facilitated, improves propulsion and manoeuvrability and compensates for weight changes as a result of different load levels and fuel and water consumption.

Prior to arrival at the port of loading, the water used as ballast is discharged into the sea and with it a multitude of organisms such as bacteria, microbes, small invertebrates, eggs, cysts and larvae of different species that once transferred to the sea can survive and establish a reproductive population in the host environment (Carlton 1985; Carlton & Geller 1993; Hewitt & Gollasch & Minchin, 2009).

The lack of control and management in these discharges has resulted in what is known as a *biological invasion* (Vila, 2008) due to the transfer of harmful aquatic organisms and pathogens from one ecosystem to another and that have resulted in the spread of invasive species that have been constituted as a threat to the marine environment, the health of human beings, goods and resources (Bax, 2003; Knowler 2005; Herbold & Moyle 2009; Davison & Simkanin 2011).

The International Community, conscious that this phenomenon poses a great threat to the conservation and sustainable use of biological diversity, promoted in 1992 the United Nations Conference on Environment and Development (UNCED 1992), which, for its part, demanded the International Maritime Organization (IMO) the adoption of measures based on globally applicable regulations with their guidelines for their effective implementation and uniform interpretation. The intention was to prevent, reduce, and finally eliminate the risk that this technology imply for the environment, as well as avoiding unwanted side effects from that control as well as to encourage developments in related knowledge to and technology.

Having regard to the legal base the provisions contained in the United Nations Convention on the Law of the Sea of 1982 relating to the protection and preservation of the marine environment and as a result of the work carried out by the Committee for the Protection of the Marine Environment, on 16 February 2004, International Convention for the Control and Management of Ships' Ballast Water and Sediments (IMO 2004) that has entered into force on September 6, 2017 (BWM Convention). Jointly, IMO has developed fourteen guidelines, essentially of a technical nature, for its effective implementation and uniform interpretation, and recently the Code for Approval of Ballast Water Management Systems (BWMS Code) has been adopted, which is expected to enter into force in October 2019 (IMO 2018).

2 EFFECTS ON MARINE ECOSYSTEMS OF BALLAST WATER

As mentioned above, the species transported together with ballast water can survive and settle a reproductive population in the host environment, becoming invasive species prevailing over native ones and proliferating to reach pest proportions.

The first scientific study echoing the problem generated by the emergence of exogenous species in other ecosystems deriving from ballast water was prepared by Ostenfeld (Ostenfeld 1908), after studying the massive emergence of Asian phytoplankton algae *Odontella*. Seventy years later, once the situation reached a worrying level, the scientific community focused its efforts on its study and possible solutions, preparing new reports that will serve as the basis for international organizations to establish a road map to follow in order to face the growing problem (Medcof 1975; Rosenthal 1980; Hallegraeff & Bloch 1992; Subba Rao 1994).

This proliferation of harmful aquatic organisms and pathogens from one ecosystem to another is largely due to the increase of international maritime traffic in recent decades. The effects can be taken into account and studied in three large blocks (García García-Revillo & Fernández-Delgado 2009):

The first one would cover the ecological impact on the area in question, resulting in a change of pre-existing biota, genetic contamination and loss of marine biodiversity.

A second blocks and direct consequence of the first one, focuses on the socio economic impact and subsequent economic losses, as would be the case of aquaculture industry and commercial fishing.

And last, but not least, the impact on public health, which is caused by the emergence of viruses and bacteria that may constitute a potential danger to human health.

In spite of the fact that the spread of invasive species is recognized worldwide as one of the greatest threats to the ecological and economic well-being of the planet, the quantitative data indicate that the number of biological invasions keeps on increasing and may not yet have reached its most serious moment. According to data from the International Maritime Organization 10,000 billion tons of ballast water are transported each year producing a new invasion every nine weeks.

In the area of the Strait of Gibraltar, the species currently classified as invasive and whose means of introduction is the water of ballast are the following: *Crassostrea gigas* or Pacific giant oyster (Daisie 2009), the tropical algae *Caulerpa taxifolia*, known as the "killer algae" (Streftaris & Zenetos 2006) and the recently discovered Asian alga *Rugulopterix okamurae* (Pulido 2017).

These species are listed in the Spanish Catalogue of Invasive Alien Species, prepared within the framework of the European Union Strategy on Invasive Alien Species, specifically in the marine environment, by means of Directive 2008/56/EC of The European Parliament and of The Council of 17 June 2008 establishing a framework for community action in the field of marine environmental policy (Marine Strategy Framework Directive).

3 STANDARDS OF MANAGEMENT, EFFECTIVENESS AND CHANGE OF BALLAST WATER

3.1 *Ballast Water Management Plan*

Each ship shall have on board and implement a Ballast Water Management Plan. Such a plan shall be approved by the Administration taking into account Guidelines developed by the Organization. The Ballast Water Management Plan shall be specific to each ship (IMO 2005).

The Ballast Water Management Plan will detail, among other things, the safety procedures for the ship and the crew related to the management of ballast water as well as the procedures for the evacuation of sediments at sea and on land (Regulation B-1).

3.2 *Ballast water management systems*

Ships shall be equipped with a ballast water treatment system whose applied technology shall be approved by the flag State in accordance with the Guidelines 8, 9, 10 and BWMS Code (IMO 2006a, 2008, 2016, 2018).

Currently the trend is the use of systems that have filtering technology to eliminate larger organisms and treatment with UV radiation to eliminate microorganisms (Eun-Chan & Jeong-Hwan & Seung-Guk 2016; Moreno-Andrés & Romero-Martínez & Acevedo-Merino & Nebot 2017; Bradie & Gianoli & He & Lo Curto 2018; Petersen, & Madsen & Glaring & Dobbs & Jørgensen 2019).

3.3 *Ballast water performance standard*

Ships that carry out the ballast water exchange will do so with an efficiency of 95% ballast water volumetric change. There must be less than 10 viable organisms per cubic meter whose minimum size is equal to or greater than 50 μ and less than 10 viable organisms per millilitre whose minimum size is less than 50 μ and greater than 10 μ; and, the discharge of the indicator microbes (*Vibrio Cholerae Toxicogeno, Escherichia Coli, Intestinal Enterococci*) will not exceed specific concentrations (Regulation D-1 and D-2).

3.4 *Ballast water exchange*

The BWM establishes as an obligation that the ballast water change be carried out at least 200 from the nearest land and in waters at least 200 metres deep. In cases where the vessel can not change the ballast water in accordance with the rule established in general, such change will be carried out as far as possible from the nearest land, and in all cases at least 50 nautical miles from the nearest land and in waters at least 200 metres deep.

In maritime areas where the distance to the nearest land or depth does not meet the parameters described, the port State, in consultation with adjacent States or with other States, as appropriate, may designate areas where the vessel effect the change of ballast water.

4 EXEMPTIONS AND RISK ASSESSMENT FOR THE PURPOSE OF ITS APPLICATION

The BWM Convention establishes in Regulation A-4 the possible exemptions that States parties may grant to ships operating in waters under their jurisdiction and provided that the following conditions are met:
1 Granted to a ship or ships on a voyage or voyages between specified ports or locations; or to a ship which operates exclusively between specified ports or locations;
2 Effective for a period of no more than five years subject to intermediate review;
3 Granted to ships that do not mix Ballast Water or Sediments other than between the ports or locations specified in paragraph 1; and
4 Granted based on the Guidelines G7 on risk assessment (IMO 2007).

In order to carry out the risk assessment for the granting of an exemption to fulfil the prescribed obligations for the management of ballast water on board, three methods are determined:

4.1 *Environmental matching risk assessment*

Environmental matching risk assessments compare environmental conditions including temperature and salinity between donor and recipient regions. These regions are typically defined as biogeographic regions and it is suggested that the Large Marine Ecosystems (LME)[1].

The degree of similarity between the locations provides an indication of the likelihood of survival and the establishment of any species transferred between those locations.

The data necessary to enable a risk assessment using environmental matching includes, but is not limited to:
1 Origin of the ballast water to be discharged in recipient port;
2 Biogeographic region of donor and recipient ports;
3 The average and range of environmental conditions, in particular salinity and temperature.

This information is used to determine the degree of environmental similarity between the donor and recipient environments.

4.2 *Species' biogeographical risk assessment*

Species' biogeographical risk assessment compares the biogeographical distributions of nonindigenous, cryptogenic, and harmful native species that presently exist in the donor and recipient ports and biogeographic regions. Overlapping species in the donor and recipient ports and regions are a direct indication that environmental conditions are sufficiently similar to allow a shared fauna and flora.

The data necessary to enable a risk assessment using a species biogeographical approach includes but may not be limited to:
1 Records of invasion in the donor and recipient biogeographic regions and ports;
2 Records of native or non-indigenous species that could be transferred through ballast water in the donor biogeographic region that have invaded Other biogeographic regions and the number and nature of biogeographic regions invaded;
3 Records of native species in the donor region that have the potential to affect human health or result in substantial ecological or economic impacts after introduction in the recipient region through ballast water transfer.

The species' biogeographical risk assessment could also be used to identify potential target species in the donor regions as indicated by native species with wide biogeographical or habitat distributions or which are known invaders in other biogeographic regions similar to that of the recipient port.

4.3 *Species-specific risk assessment*

Species-specific risk assessments use information on life history and physiological tolerances to define a species' physiological limits and thereby estimate its potential to survive or complete its life cycle in the recipient environment. That is, they compare individual species characteristics with the environmental conditions in the recipient port, to determine the likelihood of transfer and survival.

The data necessary to enable a risk assessment using the species-specific approach includes, but is not limited to:

---

[1] The Mediterranean Sea is part of the Large Marine Ecosystems according to US National Oceanic and Atmospheric Administration.

1 Biogeographic region of donor and recipient ports;
2 The presence of all non-indigenous species (including cryptogenic species) and native species in the donor ports, port region and biogeographic region, not present in the recipient port, to allow identification of target species;
3 The presence of all target species in the recipient ports, port region, and biogeographic region;
4 The difference between target species in the donor and recipient ports, port region, and biogeographic region;
5 Life history information on the target species and physiological tolerances, in particular salinity and temperature, of each life stage; and
6 Habitat type required by the target species and availability of habitat type in the recipient port.

5 CRITERIA FOR THE DESIGNATION OF AREA FOR THE EXCHANGE OF BALLAST WATER

Based on Regulation B-4, the Guidelines to be followed, although not exhaustive or exclusive, for the designation of a zone for the change of ballast water are established in the Guidelines 14 (IMO 2006b). It determines a process in which three stages are distinguished for the designation of the area: identification, evaluation and designation.

In the event that Spain considers the possibility of designating a maritime zone for the change of ballast water, it should consult with the adjacent States, that is, with Morocco and Portugal, as the case may be, when it comes to the identification, evaluation and designation of possible zones. In the event that the intention is expressed jointly, Spain together with the State or States in question will carry out a bilateral or regional agreement as appropriate for the joint establishment of the same[2].

5.1 *Identification*

According to the characteristics of the seas that surround Spain, it may be considered appropriate to identify one or more zones for the change of ballast water. In this specific case and taking as initial premises that:
1 The main routes of maritime traffic through the Strait of Gibraltar are those that are established from and towards the Bosphorus Strait, the Suez Canal, the Finisterre, the North Atlantic, the South Atlantic and the Canary Islands; and
2 The existence of commercial maritime routes between Spain and Morocco and Algeria.

That a large part of this traffic has the Port of the Bay of Algeciras, the proposal for the designation of an area for the exchange of ballast water should be directed to an area located at the eastern embouchure, that is, in the Alborán Sea, since compliance with the requirements established in Regulation B-4 and D-1 do not pose any problem for those ships that enter the Strait of Gibraltar from the western access[3].

The following considerations should be taken into account when identifying potential sea area for undertaking ballast water exchange:

5.1.1 *Legal aspects*
The provisions laid down in the International Law of the sea will be taken into consideration. Sea areas beyond the jurisdiction of a port State may provide the most practical and appropriate area for ballast water Exchange (area located more than the 200 nautical miles which constitute the exclusive economic zone). In any case you can designate a zones for the exchange of ballast water in the territorial waters of another State without its consent.

In the case of the Strait of Gibraltar, the applicable legal regime is the one established for straits used for international navigation in the United Nations Convention on the Law of the Sea, which states that the mere fact of being used as such does not affect the legal status of the waters forming such straits or the exercise by the States bordering the straits of their sovereignty.

5.1.2 *Important resources and protected areas*
It should be considered and avoid, to the extent practicable, potential adverse impact in aquatic areas protected under national or International Law, as well as other important aquatic resources including those of economic and ecological importance.

5.1.3 *Navigational constraints*
In this aspect, they will be taken into account navigation impacts, including the desirability of minimizing delays, as appropriate, taking into consideration the following: the area should be on existing routes if possible; if the area cannot be on existing routes, it should be as close as possible to them.

Constraints to safe navigation must be considered when selecting the location and size of the ballast water exchange area. Such considerations should include, but are not limited to: increased shipping traffic congestion; proximity to other vessel traffic (small craft, offshore platforms, etc.); adequate aids

---

[2] Article 13 BWM Convention

[3] Subdivision marina Strait of Gibraltar extends from West to East from the border with Portugal at the mouth of the Guadiana River to the meridian that passes through Gata Cape. The western part is the Gulf of Cádiz (Western access); the central one (between Trafalgar Cape-Espartel Cape and Punta Europa-Punta Almina) with the strait itself as a connection node between the basin of the Mediterranean Sea and the Atlantic Ocean and the eastern one, with the Alborán Sea (Eastern access).

to navigation; security of the area; and Shipping lanes/routeing systems.

One aspect to bear in mind is the type of navigation to be made by ships when they cross a strait used for international navigation in accordance with International Law of the Sea. In this case, the type of navigation is called "transit passage" and it entails the freedom of navigation solely for the purpose of continuous and expeditious transit.

5.2 *Assessment of identified sea areas*

Risk assessment (qualitative or quantitative) must be a logical process that allows determining objectively the probability and consequences of specific phenomena. The principles that should determine the characteristics and results of the risk assessment are: effectiveness, transparency, consistency, comprehensiveness, risk management, precautionary, science based and continuous improvement.

The identified ballast water exchange area should be assessed in order to ensure that its designation would minimize any threat of harm to the environment, human health, property or resources taking into account but not limited to the following factors:

5.2.1 *Oceanographic factors*
Areas where tidal flushing is poor or where a tidal stream is known to be turbid should be avoided where possible. The maximum water depth available should be selected where possible. Currents, upwelling's or eddies should be identified and considered in the evaluation process. Sea areas where currents disperse discharged ballast water away from land should be selected where possible.

5.2.2 *Physic-chemical factors*
High nutrient areas should be avoided where possible (e.g., salinity, nutrients, dissolved oxygen, chlorophyll "a").

5.2.3 *Biological factors*
Areas known to contain outbreaks, infestations, or populations of harmful aquatic organisms and pathogens (e.g., harmful algal blooms), which are likely to be taken up in ballast water, should be identified and avoided where possible.

5.2.4 *Environmental factors*
Sea area that may be impacted by pollution from human activities (e.g., areas nearby sewage outfalls) where there may be increased nutrients or where there may be human health issues should be avoided where possible. Sensitive aquatic areas should be avoided to the extent practicable.

5.2.5 *Economic factors*
Location of important resources, such as key fisheries areas and aquaculture farms should be avoided (e.g., fisheries areas, aquaculture farms).

5.2.6 *Operational factors*
A foreseen estimation of the quantities, sources and frequencies of ballast water discharges from vessels that will use the designated sea area should be considered in the assessment of such area.

5.3 *Designation of sea areas for ballast water exchange*

The location and size that provide the least risk to the aquatic environment, human health, property or resources of the southern area of Spain should be selected for designation.

In accordance with International Law, the boundaries of the area should be clearly defined. Such information should include: the precise geographical co-ordinates, depth limit and/or distance from nearest land that defines the designated ballast water exchange area; other information that may be relevant to facilitate ships' identification of the designated ballast water exchange area, for example navigation aids; and, details of the characteristics of the designated ballast water exchange area that may be relevant to assist ships plan their voyage, including: use of area by other traffic, current and tidal flow, wind and swell conditions, seasonal events (cyclones, typhoons, ice, etc.).

6 REGIONAL AGREEMENT FOR THE MANAGEMENT OF BALLAST WATER IN THE MEDITERRANEAN SEA

Part of the Mediterranean Action Plan with the technical support of the GloBallast Partnerships Project (REMPEC 2008) and taking as legal basis the obligation established in the BWM Convention in which it is determined that the States with common interests in the protection of the environment in a given geographical area and especially those that limit with closed or semi-enclosed seas, will seek to expand cooperation regional through regional agreements. To this end, the States of the Mediterranean[4] basin have drawn up a non-binding agreement that harmonizes the management of ballast water for those ships that sail in this area and are not obliged to comply with Regulation D-2 according to the schedule of

---

[4] The Contracting Parties to the Convention for the Protection of the Marine Environment and the Coastal Region of the Mediterranean (Barcelona Convention) are the following: Albania, Algeria, Bosnia and Herzegovina, Croatia, Cyprus, Egypt, The European Community, France, Greece, Israel, Italy, Lebanon Libya, Malta, Morocco, Montenegro, Slovenia, Spain, Syria, Tunisia and Turkey.

implementation established in Regulation B-3 (IMO 2011).

The practice established and in accordance with the provisions of the BWM Convention is as follows:

1 Ships entering or leaving the waters of the Mediterranean Sea area from or to the Atlantic Ocean (Straits of Gibraltar), or from or to the Indian Ocean through the Red Sea (Suez Canal): before entering or after leaving the Mediterranean Sea area; at least 200 nautical miles from the nearest land and in waters at least 200 meters in depth.

2 Ships in situations not allowing for ballast water exchange as described in (1) above, e.g. to avoid delays or deviations from a ship's intended voyage or for safety reasons: before entering or after leaving the Mediterranean Sea area; as far from the nearest land as possible, and in all cases in waters at least 50 nautical miles from the nearest land and in waters of at least 200 meters depth.

3 Ships engaged in traffic between ports located within the Mediterranean Sea area; or a port located in the Black Sea area and a port located in the Red Sea area; or a port located in the Black Sea area and a port located in the Mediterranean Sea area; or a port located in the Red Sea area and a port located in the Mediterranean Sea area: as far from the nearest land as possible, and in all cases in waters at least 50 nautical miles from the nearest land and in waters of at least 200 meters depth.

4 Ships in situations not allowing for ballast water exchange as described in (3) above, e.g. to avoid delays or deviations from a ship's intended voyage or for safety reasons: areas designated by the port State for that purpose.

7 REGIONAL CHARACTERISTICS OF THE STRAIT OF GIBRALTAR FOR THE PURPOSES OF DESIGNATING A BALLAST WATER ZONE

The main factors to be taken into account for determining a ballast water zone in the Strait of Gibraltar are:

1 In the Alboran Sea, Spain's territorial sea and contiguous zone is delimited, unlike Morocco that has established an exclusive economic zone;

2 The existence of two large fisheries (Gulf of Cádiz and the Mediterranean Sea). Fishing is considered a strategic industry because of its financial contribution to the region of Andalusia. However, it is in continuous decline due to its overexploitation (Robles 2007);

3 An area rich in biological diversity. Particularly noteworthy are the marine reserves of the Seco de

los Olivos, Gata-Nijár and the Island of Alboran (MARINEPLAN 2007);

4 International navigation strait. The Port of Algeciras registers a high volume of marine traffic[5]; and,

5 Existence of traffic separation schemes: the Alboran Sea is crossed by the route from the Mediterranean Sea to the Atlantic and vice versa. Ships access the Alboran Sea on its eastern side through the traffic separation scheme located at Cabo de Gata and then exit to the Atlantic Ocean via a traffic separation scheme located in the Strait of Gibraltar itself.

8 CONCLUSIONS

In line with the global concern about the deterioration of the planet, and aiming clear protectionists goals, the BWM Convention contains in its provisions strict measures to counteract the effects of the water of ballast on marine ecosystems, paying special attention to those geographical areas in which the general operational limits can not be carried out.

This usually occurs in straits or enclosed seas with geographical features that do not allow compliance with the minimum distance from the coast of 50 miles for the discharge of ballast water.

As a solution, the possibility has been raised of establishing specific areas for the discharge of ballast water through bilateral or regional agreements, or unilaterally granting exemptions to vessels operating in waters under a single jurisdiction depending on the specific features of the ports of origin and destination and a prior assessment of environmental and biogeographical risks.

In the case of the Strait of Gibraltar, the problem arises specifically with respect to the marine traffic that runs between Spain and Morocco in the eastern part via the Alboran Sea.

By means of a Spanish initiative and on the proposal of the Association of Spanish Shipping Companies (ANAVE), talks have started to jointly establish an intermediate zone for the discharge of ballast water. However, the designation procedure is expected to be slow and complicated, given the lack of interest on the part of the Moroccan authorities and the shipping companies in providing specific data on ballast water to facilitate a thorough assessment of actual needs.

Along with this, and in compliance with IMO guidelines, there is the further problem posed by the legal, economic, environmental and operational factors present in the Alboran Sea and which would

---

[5] In 2018, the Port of Algeciras registered the entry of 2,115 vessels accounting for 30,909,71 GT. http://www.apba.es/estadisticas (Accessed 5 March 2019)

complicate the designation of an intermediate ballast water exchange zone between Spain and Morocco. These can be summarised as follows:

1 Jurisdictional difference regarding the delimitation of marine spaces;
2 Major marine and fishing reserves;
3 International status as a strait and restrictions regarding safety of navigation.

REFERENCES

Bax, N. 2003. Marine invasive alien species: a threat to global biodiversity. *Marine Policy* 27: 314-315.

Bradie, J. & Gianoli, C. & He, J. & Lo Curto, A. 2018. Detection of UV-treatment effects on plankton by rapid analytic tools for ballast water compliance monitoring immediately following treatment. *Journal of Sea Research* 133: 177-184.

Carlton J. T. 1985. Transoceanic and interoceanic dispersal of coastal marine organisms: the biology of ballast water. *Oceanogr Mar Biol Ann Rev* 23:313-371

Carlton, J. T. & Geller, J.B. 1993. Ecological roulette: the global transport of nonindigenous marine organisms. *Science* 261(2): 78-82.

DAISIE. 2009. *Handbook of Alien Species in Europe*. In J. A. Drake (eds.). Berlin: Springer.

Davinson, I. C. & SimKanin, C. 2011. The biology of ballast water 25 years later. *Biol Invasions* 14: 9-13.

Eun-Chan, K. & Jeong-Hwan, O. & Seung-Guk L. 2016. Consideration on the Maximum Allowable Dosage of Active Substances Produced by Ballast Water Management System Using Electrolysis. *International Journal of e-Navigation and Maritime Economy* 4: 88-96.

García García-Revillo & M., Fernández-Delgado, C (ed.) 2009. *La introducción por mar de especies exóticas invasoras a través del agua de lastre de los barcos. El caso de Doñana*: 25. Córdoba: Universidad de Córdoba.

Hallegraeff, G. M. & Bloch, C. J. 1992. Transport of diatom and dinoflagellate resting spores in ship´s ballast water: implication for plankton biogeography an aquaculture. *J. Plankton Res* 14 (8): 1067-1084.

Herbold, B. & Moyle, P. B. 1986. Introduced species and vacant niches. *Am Nat* 128: 751-760.

Hewitt, C. L. & Gollasch, S. & Minchin, D. 2009. The vessel as a vector–biofouling, ballast water and sediments. In Rilov G, Crooks JA (eds.) *Biological invasions in marine ecosystems*: 117–131. Berlin: Springer.

IMO. International Maritime Organization. 2004. International Convention for the Control and Management of Ships' Ballast Water and Sediments, BWM/CONF/36.

IMO. International Maritime Organization. 2005. Resolution MEPC.127 (53). Guidelines for ballast water management and development of Ballast Water Management Plans (G4).

IMO. International Maritime Organization. 2006b. Resolution MEPC.151 (55), Guidelines on designation of areas for ballast water exchange (G14).

IMO. International Maritime Organization. 2006a. Resolution MEPC.140 (54). Guidelines for approval and oversight of prototype ballast water treatment technology programmes (G10).

IMO. International Maritime Organization. 2008. Resolution MEPC.169 (57). Procedure for approval of ballast water

management systems that make use of active substances (G9).

IMO. International Maritime Organization. 2011. Harmonized voluntary arrangements for ballast water management in the Mediterranean Region, BWM.2/Circ.35. Annex 1.

IMO. International Maritime Organization. 2016. Resolution MEPC.278 (70). Guidelines for approval of ballast water management systems (G8).

IMO. International Maritime Organization. 2017. Resolution MEPC.289 (71), Guidelines for risk assessment under regulation A-4 of the BWM Convention (G7).

IMO. International Maritime Organization. 2018. Resolution MEPC.300 (72). Code for Approval of Ballast Water Management Systems (BWMS CODE).

Knowler, D. 2005. Reassessing the costs of biological invasion: Mnemiopsis leidyi in the Black Sea. *Ecological Economics* 52: 187-199.

MARINEPLAN. 2007. *La Política Marítima y planificación espacial. Aplicación metodológica arco atlántico-mediterráneo (Golfo de Cádiz y mar de Alborán)*. Proyecto MEC (SEJ 2007-66487/GEOG). INFORME 6. Caracterización de la Subdivisión "Estrecho" pp. 20-23.

Medcof, J. C. 1975. Living Marine animals in Ships´ Ballast water. *Proc. Natl. Shellfish* 2: 321-322.

Moreno-Andrés, J. & Romero-Martínez, L. & Acevedo-Merino, A. & Nebot, E. 2017. UV-based technologies for marine water disinfection and the application to ballast water: Does salinity interfere with disinfection processes?. *Science of The Total Environment* 581–582: 144-152.

Ostenfeld, C. H. 1908. On the Immigration of Biddulphia sinensis Grev and its Occurrence in the North Sea during 1903-1907. *Medd. Komm. Havunders. Ser. Plankton* 1(6): 1- 46.

Petersen, N. B. & Madsen, T. & Glaring, M. A. & Dobbs, F. C. & Jørgensen, N. O. G. 2019. Ballast water treatment and bacteria: Analysis of bacterial activity and diversity after treatment of simulated ballast water by electrochlorination and UV exposure. *Science of The Total Environment* 648: 408-421.

Pulido Leire, C. 2017. *Rugulopteryx okamurae (Dictyotales, Ochrophyta): Morfología, anatomía y estrategias reproductoras de una nueva especie exótica de macroalga en el Estrecho de Gibraltar*. Málaga: Universidad de Málaga.

Robles, R. 2007. Conservación y desarrollo sostenible del Mar de Alborán: elementos estratégicos para su futuro gestión. Centro de Cooperación para el Mediterráneo. Unión Mundial para la Naturaleza.

REMPEC. The Regional Marine Pollution Emergency Response Centre for the Mediterranean Sea. 2008. GEF/UNDP/IMO Project "Building partnerships to assist developing countries to reduce the transfer of harmful aquatic organisms in ship's ballast water (GloBallast Partnerships)".

Rosenthal, H. 1980. Implications of transplantations to aquaculture and ecosystems. *Mar, Fish. Rev* 5:1-14.

Streftaris, N. & Zenetos, A. 2006. Alien Marine Species in the Mediterranean - the 100 Worst Invasives and their Impact. *Mediterranean Marine Science*, 7(1): 87-117.

Subba Rao, D.V. 1994. Exotic phytoplankton From ship´s ballast water: risk of potential spread to mariculture sites on Canada's east coast. *Can Data Rep. Fish. Aquatic Sci* 937: 1-51.

UNCED. United Nations Conference on Environment and Development. 1992. The Rio Declaration on Environment and Development, A/CONF.151/26 (Vol. I) Principle 1 and 2.

Theoretical Research on Mass Exchange Between an Autonomous Transport Module and the Environment in the Process of Transport from the Seabed

W. Filipek & K. Broda
AGH University of Science and Technology, Krakow, Poland

ABSTRACT: In recent years, we have witnessed a substantial increase in the interest in maritime mining in the world, which has resulted in the development of many new methods for the exploitation of marine deposits as well as transport of spoil from the seabed to the surface (Karlic, 1984, Depowski, Kotliński, Rühle, Szamałek, 1998, Abramowski and Kotliński, 2011, SPC, 2013, Royal Society, 2017, Sharma, 2017, Jones et al, 2019, Websites: JOGMEC, Nautilus Minerals, Atlantis II Deep, Bluenodules). In their research, the authors have been working on the development of a new concept of transport from the seabed.

The authors have already presented the concept of the principle of operation of an autonomous transport module for transport from the seabed (Filipek and Broda, 2016, 2017, 2018). They designed and constructed a laboratory stand and carried out experiments, the results of which confirmed the applicability of the concept discussed in transport from the seabed. The research also included the determination of the energy source in the transport process and the changes in the average density of the transport module. Three transport concepts from the seabed were also compared in terms of their energy demand. In the next step, the authors addressed the issues of stability of the transport module in the process of immersion and ascent.

In the course of these studies, there appeared the necessity of the theoretical examination of mass transfer between the transport module and the environment, which is the subject of this publication. The in-depth analysis of this issue is essential from the point of view of the practical suitability of the module for transporting from the seabed because the course of mass exchange between the transport module and the environment determines the speed of immersion and ascent.

1 INTRODUCTION

In their previous studies (Filipek and Broda, 2016, 2017, 2018), the authors did not present any (even theoretical) concept of storing reaction products which undergo phase transitions, and which are responsible for the emergence process. So far, we have assumed excess (gaseous or liquid) products to be released into the environment. While, thus far, we have focused on the issue of initiating the process, we have always paid attention to the economic use of excess energy and also to minimising the environmental impact.

In this article, we seek to present a theoretical concept of solving this issue, bearing in mind that it is far from perfect. We based this concept on the operating model of an autonomous transport module. At this point, we would like to underline that this proposal can also be based on other gaseous media – especially hydrogen. For the simplicity of determining one of the directions of our research, we have decided to analyse our concept using the simplest example, in which acetylene is used as the gaseous medium. The physical and chemical changes which take place in the module are easily verifiable with simple laboratory methods (Filipek and Broda, 2018). The experience gained in developing the concept in question will enable us to model more complex processes, using other gaseous media, constituting the source of energy in the process of transporting the excavated material from the seabed.

Pure acetylene (ethyne) C_2H_2 is a colourless, odourless and lighter-than-air gas with a density of $\rho=0.0010996$ g/cm³ (15 °C, 1000 hPa). As a technical product, it has an unpleasant odour. The acetylene obtained through the hydrolysis of calcium carbide, the pyrolysis of methane or halogen elimination is stored (once dissolved in acetone) in steel cylinders under a pressure of 1.2 MPa (Mizerski 2013,Urbański, 1947; https://pubchem. ncbi.nlm.nih.gov/ compound/6326). Since the molecule of acetylene has a high percentage content of carbon (92.3%), it is not subject to complete combustion in air (21% of oxygen).

Acetylene-air mixtures are explosive if they contain from 2.4% to 83% of acetylene. Due to an external stimulus, such as a rapid local increase in pressure, temperature or electrical discharge, it can rapidly, and in an explosive manner, decompose into carbon and hydrogen, releasing a significant amount of energy, ΔH_1= 228.2 kJ/mol, in accordance with the following formula:

$$C_2H_2 \; 2C + H_2 + H_1 \tag{1}$$

This phenomenon is used, on an industrial scale, in a controlled process of acetylene decomposition into hydrogen and carbon, which is carried out under a pressure increased to 4-6 bar, at low temperatures, in order to obtain the so-called acetylene soot. The electrical discharge is used as an impulse/stimulus triggering acetylene decomposition.

When the acetylene decomposition temperature reaches 1000 °C, the hydrogen H_2 molecule decomposes into atoms, in accordance with the following formula (Mizerski, 2013):

$$H_2 + H_2 \xrightarrow{1000°C} 2H + H_2 \tag{2}$$

It simultaneously absorbs the activation energy, E_a= 402 kJ/mol. One can, therefore, conclude that, in the course of rapid decomposition, a significant proportion of hydrogen will decompose into hydrogen atoms H which will spontaneously seek to become bound into a hydrogen molecule H_2, thus resulting in sudden spikes of pressure.

Furthermore, acetylene is widely used in the chemical industry (e.g. in the production of polymers), as well as in welding and medicine – for general anaesthesia (the so-called narcylene). In the past, it was used in acetylene gas lamps for lighting purposes. It is not a "physically suffocating" gas. However, if its concentration exceeds 50%, it can cause fatal asphyxiation. Usually, before its concentration reaches harmful levels, a rapid and explosive reaction with oxygen takes place.

The literature does not clearly specify the physical conditions (and especially pressure values) under which one can expect the spontaneous decomposition of acetylene to occur. The authors' experience (Filipek and Broda, 2018) indicates that acetylene appears to be stable under pressure values close to, or slightly exceeding, the atmospheric pressure. On the other hand, in the course of decompression of pressurised acetylene to the atmospheric pressure, we observed a substantial electrification of the environment, persisting for quite a long time. Despite using small amounts of calcium carbide of approx. 500 mg, which corresponded to generating approx. 200 cm^3 of acetylene, (Filipek and Broda, 2018) in a well-ventilated room with a cubic capacity of approx. 70 m^3, clear electrification of objects took place. This caused the observed electrical discharges and might

be the main stimulus contributing to the initiation of acetylene decomposition. As presumed by the authors, acetylene itself does not decompose spontaneously, and there must be a stimulus resulting from physical properties of the newly created substances originating in the compression or decompression of acetylene.

We have selected calcium acetylide (carbide) CaC_2 as the source of energy. It is commonly used in the industrial method of acetylene production (Morrison and Boyd, 1997; Kaczyński and Czaplicki, 1977). Calcium acetylide is produced through a reaction of calcium oxide and coke at a very high temperature.

Calcium acetylide CaC_2 reacts with water H_2O, by means of reaction (3) well documented in the literature (Kaczyński and Czaplicki, 1977), producing acetylene C_2H_2 and calcium hydroxide $Ca(OH)_2$, and generating a substantial amount of energy ΔH_2 = -123.24 kJ/mol.

$$CaC_{2(s)} + 2H_2O_{(c)} \rightarrow C_2H_{2(g)} + Ca(OH)_{2(s)} + \Delta H_2 \tag{3}$$

In order to intensify the decomposition process, in our research calcium carbide was submerged in an inactive, but well water-soluble, substance (Filipek and Broda, 2018).

2 THE DETERMINATION OF DEMAND FOR ACETYLENE

NH_4NO_3 The volume of working gas (in our case acetylene) produced in the decomposition or phase transition process is strictly dependent on three parameters, i.e. the number of moles n; temperature T and pressure p. The values of pressure, temperature and density of water, as a function of depth, which we used in our calculations, are based on the data obtained from the EX1504L2 measurement point, downloaded from the Ocean Exploration and Research website (https://www.ncddc.noaa.gov/website/google_maps/OE/mapsOE.htm 6.12.2018). That way, we wanted to refer in our research to the real conditions found in the ocean, rather than limiting ourselves to purely theoretical considerations. The calculation results are presented in Figure 1. The van der Waals equation was used in our calculations. Its use for determining the parameters of state in our deliberations is justified in our previous article (Filipek and Broda, 2018).

The curves shown in Fig.1 illustrate the relationship between the number of moles of the gaseous medium in question (in our case C_2H_2) and the submersion depth of the transport module, and thus the pressure occurring at this depth.

The three curves, T_0, T_K, T_{50} (Fig.1), were defined using the van der Waals equation (Berry and

Rice and Ross, 2000), and they illustrate, in quantitative terms, the demand for moles n of the gaseous medium, so that its volume is 1 m³ under pressure at a given depth. Adopting the volume of 1 m³ allows us, when keeping the appropriate proportions, to determine the number of moles n of working gas for any volume.

Curve T_0 represents a quasi-static course characterised by the working gas temperature's being equal to water temperature at a given depth.

As regards curve T_K, we assumed that, in the process in question, the temperature of the medium was constant and equal to the critical point. On the other hand, curve T_{50} refers to the case in which the working gas temperature is assumed at 50 °C throughout the entire process. Considering that, in the analysed case, the working gas formulation according to formula (3) is accompanied by a significant release of energy in the form of heat, curve T_0 can be viewed as the limit curve below which every actual reaction course will be observed. On the other hand, since we can assume, with a high degree of probability, an intensive heat exchange with the environment (e.g. the steel structure of the module) and a long submersion time, the temperature of the gaseous medium should not rise significantly above 50 °C. Since the acetylene medium considered in our article undergoes a phase transition under certain pressure and temperature conditions, we obtain a vertical and rapid increase in the number of moles n at the point of discontinuity corresponding to the phase transition conditions, which can be observed for curve T_0. This phenomenon will take place until T_K. Since the energy balance was disregarded in our deliberations (it will be the subject of future research), we decided to consider all processes as being quasi-static. It is worth recalling that the curves discussed above illustrate the demand for the working gas amount in the submersion process, so that it has a volume of 1 m³ at the pressure and temperature occurring at a given depth.

In the next step, we will deal with modelling the process of working gas production according to reaction (3) so that the final course cuts through the aforementioned curves at a point corresponding to the selected depth, i.e. so that the needed amount of the gaseous medium equals the generated amount.

The linear function is the baseline limit curve (Fig.1)

$$n = aH \tag{4}$$

where coefficient a was determined using formula (5)

$$a = \frac{\partial n_1(0)}{\partial H(0)} \tag{5}$$

For acetylene selected as the working gas, a=3.974 [mol/m].

When modelling a case where the desired depth is to be greater than it is possible according to the limit function, the amount of working gas, as a function of depth, can be presented in the form of the following formula (Fig.1).

$$n = a(1-b)H \tag{6}$$

where $b \in R_+$.

Where the desired depth is lower than the limit, the demand for the working gas can be modelled using the function (Fig.1)

$$n = aHe^{cH} \tag{7}$$

where $c \in R_+$.

When we want the amount of gas generated in the initial period of submersion to be below the limit curve, $n = aH$, it will constitute a special case. This case can be modelled using the function (Fig.1)

$$n = a(1-d)He^{cH} \tag{8}$$

where $d \in R$.

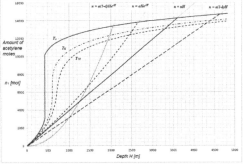

Figure 1. Demand for the working medium.

With such assumptions, it is guaranteed that the autonomous transport module will submerge to a set depth. This is due to the fact that, from the start of the submersion process to reaching the planned depth, the demand for the working gas will be lower than the amount of the generated gas.

3 THE THERMODYNAMIC CYCLE OF THE WORKING MEDIUM (ACETYLENE) IN AN AUTONOMOUS TRANSPORT MODULE, CONDITIONING THE SUBMERSION AND EMERGENCE PROCESSES.

Focusing on density in the aforementioned deliberations, one can note that the immersion process will be conditional on the module meeting the negative buoyancy requirement:

$$\rho_{H_2O}(H) \le \frac{m_1(H)+m_2+m_3+m_4}{V_1(H)+V_2+V_3+V_4} \qquad (9)$$

where indices 1,2,3,4 mean acetylene C_2H_2, calcium carbide CaC_2, calcium hydroxide $Ca(OH)_2$ and acetone C_3H_6O, respectively; m – masses; V – volumes; ρ - densities. Because of the state of matter, the values m_1, ρ_{H_2O}, ρ_1, V_1 depend on depth H, while the remaining values are constant. After a simple transformation, formula (9) can read:

$$m_1(H)\left(\frac{\rho_{H_2O}(H)}{\rho_1(H)}-1\right) \le m_2\left(1-\frac{\rho_{H_2O}(H)}{\rho_2}\right)+m_3\left(1-\frac{\rho_{H_2O}(H)}{\rho_3}\right)+m_4\left(1-\frac{\rho_{H_2O}(H)}{\rho_4}\right) \qquad (10)$$

By introducing the number of moles n and molar mass M in accordance with the aforementioned indices, while also taking into account the fact that the density of acetone ρ_4 is lower than that of water, we obtain:

$$n_1(H)M_1\left(\frac{\rho_{H_2O}(H)}{\rho_1(H)}-1\right) \le (n_2-n_1(H))M_2\left(1-\frac{\rho_{H_2O}(H)}{\rho_2}\right)+ \qquad (11)$$
$$n_3M_3\left(1-\frac{\rho_{H_2O}(H)}{\rho_3}\right)-n_4M_4\left(\frac{\rho_{H_2O}(H)}{\rho_4}-1\right)$$

By introducing the following quantifications:

$$A(H) = M_1\left(\frac{\rho_{H_2O}(H)}{\rho_1(H)}-1\right)$$

$$B(H) = M_2\left(1-\frac{\rho_{H_2O}(H)}{\rho_2}\right)$$

and

$$C(H) = M_3\left(1-\frac{\rho_{H_2O}(H)}{\rho_3}\right)$$

$$D(H) = M_4\left(\frac{\rho_{H_2O}(H)}{\rho_4}-1\right)$$

we obtain:

$$n_1(H)A(H) \le (n_2-n_1(H))B(H)+n_3C(H)-n_4D(H) \qquad (12)$$

An interesting, and perhaps the most important, parameter in the submersion process is the maximum number of moles of acetylene $n_1(H)$ produced, given the assumed values m_2, m_3, m_4 (i.e. n_2, n_3, n_4), in order for the submersion process to occur. One can determine this value using formula (12)

$$n_1(H) \le \frac{n_2B(H)-n_4D(H)}{A(H)+B(H)-C(H)} \qquad (13)$$

In the formula presented above, n_4 may not take a random value. The limit value of n_4, which will remain constant during submersion, can be determined for the starting point H=0, given that

there is an initial amount of acetylene moles n_1, e.g. from the previous cycle, and that the new cycle begins with some values n_2 and n_4 being taken into account. On the other hand, due to the new process not being initiated yet, $n_3=0$ and the intended value of n_4 is determined by the relationship presented in formula (12)

$$n_4 = \frac{n_2B(0)-n_1(0)\left[B(0)-A(0)\right]}{D(0)} \qquad (14)$$

Now, let us analyse a model transport cycle consisting of both submersion and emergence. We should bear in mind that the baseline for our deliberations is the assumption that at point A (Fig.2), upon starting the emergence process, the volume of the working gas (acetylene) is 1 m^3 and n_2 specifies the number of moles of calcium carbide needed to generate the amount of acetylene assumed at point A. In order to better illustrate these phenomena, limit curve T_0, whose significance is presented above in relation to Fig.1, was plotted in Fig.2. In order to clearly present the processes, coefficient α, which provides information about the change in the number of moles of acetone, in relation to the limit value n_4, was introduced to formula (13). Then, formula (13) reads as follows:

$$n_1(H) \le \frac{n_2B(H)-\alpha n_4D(H)}{A(H)+B(H)-C(H)} \qquad (15)$$

Assuming that $\alpha=0$, i.e. that we omit the element related to the weight of acetone in formula (15), we obtain the course illustrated in Fig.2 by the limit curve, $\alpha=0$. This course determines the limit value of the demand for acetylene, to balance the weight of the products of calcium carbide decomposition (into calcium hydroxide). By analysing this situation, we are likely to arrive at a conclusion that the maximum mass of excavated material m_5 (16), which can be transported from depth H, corresponding to point A, amounts to approx. 16.3 % of the initial amount of the working substance (calcium acetylene).

$$m_5 \le \frac{n_1(H)A(H)-n_3C(H)+n_4D(H)}{\left(1-\frac{\rho_{H_2O}(H)}{\rho_5}\right)} \qquad (16)$$

This unsatisfactory amount results from the necessity to transport to the surface a significant amount of the solid products of calcium carbide degradation. By introducing acetone, whose density is lower than that of the surrounding water, we can extract more excavated material from the seabed. The application of acetone results in the limit curve for the adopted α assuming lower values than for curve $\alpha=0$ (Fig.2). As it can be observed, when approaching a depth corresponding to straight line

AB, the maximum amount of acetone moles which can be generated during submersion decreases along with an increase in α. This is why at this depth it is necessary to wait, generating the missing amount of acetylene, and moving along straight line BA until point A is reached. Then, the emergence process is launched. Coefficient α, however, may not be increased endlessly because, as it can be seen in Fig. 2, in the case of, e.g., α=0.75 there is a possibility that the amount of acetylene generated at a depth of approx. 2000 m, equal to the permissible limit value for this depth, will prevent further submersion at greater depths, given the decrease in the required amount.

On the other hand, for α=0.9 it is impossible to reach a depth corresponding to straight line AB, which is presented in Fig.2.

Let us assume that during submersion we intend to reach point B corresponding to the course for α=0.75. During submersion, the amount of generated acetylene can be expressed (as explained while discussing Fig.1) using straight line CB, where $n_1=n_1(0)$ at point C (determined according to formula (15)), bearing in mind that, in the submersion process, we take the amount of calcium carbide which is sufficient to reach point A after reaching point B. With such a cycle, the excavated material which can be transported from the seabed to the surface equals 103% of the original mass of calcium carbide.

Starting from point C, i.e. upon commencing submersion and reaching the anticipated depth corresponding to point B, it is necessary to generate – while staying at this depth – the amount of the gaseous medium indispensable for reaching point A. The excavated material is loaded at this time. After reaching point A, the emergence process is launched (section AC in Fig.2). In order for the emergence process to continue, the minimum amount of acetylene in the system must not be lower at a given depth than the amount specified by curve W.

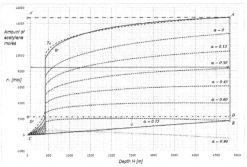

Figure 2. The cycle of the working medium carried out by the autonomous transport module.

As presented in Fig. 2, curve W does not correspond perfectly with curve T_0. At the early stage of emergence, curve W lies slightly above curve T_0, and then crosses it and runs below until the complete emergence at point C. During emergence, the working medium is decompressed, meaning that its volume increases, which is unfavourable for the operation of the autonomous transport module.

In our research, we have developed two (theoretical) concepts for solving this problem. One consists in absorbing the excess working gas generated at point A by an absorbent. In our deliberations, acetylene acts as the working gas, which is very well absorbed by acetone. As acetone is less dense than water, in addition to absorbing acetylene, it also plays a crucial role in determining the maximum mass of the extracted material that can be taken from the seabed during emergence, which is discussed above in relation to determining coefficient α and its role in the operation of the device. We should point out that it is possible to store in the autonomous transport module more acetone than it would follow from formula (15). This is due to the fact that its density is lower than the density of seawater, so its mass can be used to balance the mass of the structure of the autonomous transport module to achieve the required density of the object in question.

The other concept assumes that, as long as the working medium is liquid, it will be not absorbed in acetone, but within a system of two or more cooperating transport modules. Any excessive amount of the working medium generated during emergence is transported to a module (modules) on the seabed. This means that the excessive amount of the working medium (section A'D' in Fig.2) can be reused (section AD). Under this concept, the excessive amount of the working medium in the gaseous phase (section D'C) is absorbed in acetone. With such a cycle, the excavated material which can be transported to the surface equals to approx. 350% of the original mass of calcium carbide.

As this is a complex issue, we will devote more space to these concepts in other articles.

4 CONCLUSIONS

In order to determine the mass transfer between an autonomous transport module and the environment, in the process of transporting excavated material from the seabed, it was first necessary to determine the demand for acetylene. This is discussed in detail above. As already noted, our deliberations were based on the premise that the number of moles n_1 of the gaseous medium at point A must ensure that its volume under pressure at a given depth equals 1 m³. Knowing the demand for moles as a function of depth (curve T_0 in Fig.1 and Fig.2), we can determine the volume of the generated working medium at any point of the process in question, by applying the appropriate proportions. Knowing the volume of the gaseous medium inside the module in question, we can determine the volume, and thus the mass transfer between the module and the environment. The course of such transfer is presented in Fig.3 in respect of the conditions specified for cycle CBAC as shown in Fig. 2. The analysis of mass transfer illustrated in Fig. 2 (cycle CBAC) clearly points to the need for defining point C' in Fig.3. This point should not be plotted in Fig. 2, as doing so would render the whole result unclear, due to a very short distance to point C, using the scale adopted for Fig. 2. Section CC' precisely corresponds to the mass of the extracted material transported from the bottom to the surface.

As regards the working medium (acetylene), in points C and C' it displays the same temperatures and pressures, with the difference lying in the number of moles, and thus in the volume and mass. Obviously, during the decomposition of the working medium, it must be absorbed in acetone. This results in the gaseous medium occupying a smaller volume, which is, in turn, occupied by water. Point C', therefore, approaches point C and, hence, the mass transfer cycle between water the environment closes.

In Fig.3 curve H illustrates the demand for water to generate the amount of acetylene needed at a given depth, through reacting with calcium carbide (3). After reaching the set depth (point E), the reaction generating the working medium continues, which is why the demand for water will grow until point F is reached.

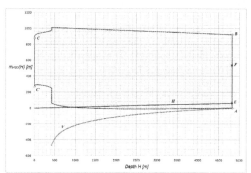

Figure 3. Mass transfer during the cycle of the working medium carried out by the autonomous transport module.

If no working medium is absorbed in the emergence process (launched at point A, Fig.3), the working medium will have to expand, given the constant number of moles of the working medium in the discussed module, as well as changes in the pressure of the environment, thus increasing its volume and pushing the relevant mass of water outside the transport module, which is illustrated by

curve V in Fig.3, determined for a liquid working medium.

An absorbent of the working medium, when used in the autonomous transport module, offers a range of advantages. The most important one concerns the possibility of regulating the amount of the working medium, in particular in the emergence process. The acetylene-acetone system makes this possible. Furthermore, due to acetone's being less dense than water, it can balance the weight of the products of the reaction generating the working medium, which improves the displacement of the autonomous transport module. In addition, it is not insignificant that acetylene is distributed in the form of acetylene dissolved in acetone. The demand for acetylene facilitates the sales of its excessive amounts resulting from this module. The possibility of utilising the working medium in the subsequent cycles enables a substantial reduction of the amount of calcium carbide used as the source of the working medium, while retaining the same lift capacity of the autonomous transport module. Theoretical studies conducted to date have shown that we are able to reuse the entire amount of the working medium in the liquid phase (sections AD or A'D' in Fig.3). We are currently conducting research into lowering limit DD', to shift it towards the working medium in the gaseous phase, to recover additional amounts of this medium to be used in the subsequent cycles, as illustrated by the 'arrow' in Fig.2.

Despite the aforementioned advantages, we are aware of the unstable nature of acetylene, especially in the liquid phase, which is why we are also researching other absorbate-absorbent systems. Ideally, the working medium should be selected in such a way that, following phase transition, the total mass of substrates would be transformed into the mass of the working medium, or their density would be lower than the density of seawater. If so, there would be no need in the emergence process to transport products which, as evidenced on the basis of the acetylene-acetone system, constitute an unfavourable ballast in the density balance formula for the autonomous transport module.

ACKNOWLEDGMENTS

This article was written within Statutes Research AGH, No. 16.16.100.215

REFERENCES

Abramowski, T. & Kotliński, R. 2011. Współczesne wyzwania eksploatacji oceanicznych kopalin polimetalicznych. Górnictwo i geoinżynieria. Rok 35, zeszyt 5, pp.41-61.
Berry, R.S. & Rice, S.A. & Ross J. 2000. Physical chemistry, Oxford University Press, pp 298-306

Depowski, S. & Kotliński, R. & Rühle, E. & Szamałek, K. 1998. Surowce mineralne mórz i oceanów, Wydawnictwo Naukowe Scholar, Warszawa

Filipek, W. & Broda, K. 2016. Theoretical foundation of the implementation of controlled pyrotechnical reactions as an energy source for transportation from the sea bed. Scientific Journals of the Maritime University of Szczecin 48 (120): 117-124

Filipek, W. & Broda, K. 2017. The Theoretical Basis of the Concept of Using the Controlled Pyrotechnical Reaction Method as an Energy Source in Transportation from the Sea Bed, TransNav the International Journal on Marine Navigation and Safety of Sea Transportation, Vol.11, No. 4, pg. 653-659.

Filipek, W. & Broda, K. 2018. Theoretical research on the gas phase density change in processes occurring during work of the transport module intended for transport from the seabed. New Trends in Production Engineering; ISSN 2545-2843; ISBN: 978-83-952420-0-7; vol. 1 iss. 1., pg. 597–604. Warszawa : STE Group Sp. z o.o.

Filipek, W. & Broda, K. 2018. Research on the concept of using calcium carbide as a source of energy for transport from the seabed. New Trends in Production Engineering ; ISSN 2545-2843; ISBN: 978-83-952420-0-7; vol. 1 iss. 1., pg. 277–284. Warszawa : STE Group Sp. z o.o.

Filipek W., Broda K. 2018, Experimental research on the concept of using an autonomous transport module for transport from the seabed. New Trends in Production Engineering; ISSN 2545-2843; ISBN: 978-83-952420-0-7; vol. 1 iss. 1., pg. 267–275. Warszawa: STE Group Sp. z o.o.

http://www.blue-nodules.eu/ [December 2018]

https://www.manafai.com/atlantisiideep.php [January 2019]

http://www.jogmec.go.jp/english/stockpiling/metal_10_000002.html [January 2019]

http://www.nautilusminerals.com [12 March 2016]

https://www.ncddc.noaa.gov/website/google_maps/OE/mapsOE.htm [6 December 2018]

https://pubchem.ncbi.nlm.nih.gov/ compound/6326 [8 February 2018]

Jones D.O.B., Durden J.M, Murphy K., Gjerde K.M., Gebicka A., Colaço A., Morato T., Cuvelier D., Billett D.S.M. 2019. Existing environmental management approaches relevant to deep-sea mining. Marine Policy (article in press) Available from: https://www.sciencedirect.com/science/article/pii/S0308597X18303956 [February 2019]

Kaczyński J., Czaplicki A. 1977, Chemia ogólna, Wydawnictwo Naukowo-Techniczne, Warszawa.

Karlic, S. 1984. Zarys górnictwa morskiego, Wydawnictwo „Śląsk", Katowice

Mizerski, W. 2013. Tablice chemiczne, Grupa Wydawnicza Adamantan s.c., Warszawa

Morrison, R.T. & Boyd, R.N. 1997. Chemia organiczna, Wydawnictwo naukowe PWN, Warszawa

The Royal Society. 2017. Future Ocean resources: Metal-rich minerals and genetics – evidence pack. ISBN 978-1-78252-260-7. Available from: https://royalsociety.org/~/media/policy/projects/future-oceans-resources/future-of-oceans-evidence-pack.pdf [5 January 2019]

Sharma, R. 2017. Deep-Sea Mining: Resource Potential, Technical and Environmental Considerations, Springer International Publishing AG, Cham

SPC, 2013 Deep Sea Minerals: Sea-Floor Massive Sulphides, a physical, biological, environmental, and technical review. Baker, E., and Beaudoin, Y. (Eds.) Vol. 1A, Secretariat of the Pacific Community, ISBN 978-82-7701-119-6, Available from: http://gsd.spc.int/dsm/public/files/meetings/Train-ingWorkshop4/UNEP_vol1A.pdf [12 July 2015]

Urbański, T. 1985. Chemistry and Technology of Explosives. Oxford: Pergamon Press

Author index

Printed and bound by CPI Group (UK) Ltd, Croydon, CR0 4YY

24/10/2024

01778293-0017